ANALOG DEVICES TECHNICAL REFERENCE BOOKS

<u>Published by Prentice Hall</u>
Analog-Digital Conversion Handbook
Digital Signal Processing in VLSI
Digital Signal Processing Applications Using the ADSP-2100 Family
Digital Signal Processing Laboratory Using the ADSP-2101 Microcomputer

<u>Published by Analog Devices</u>
Nonlinear Circuits Handbook
Transducer Interfacing Handbook
Synchro & Resolver Conversion
High-Speed Design Seminar
Mixed-Signal Design Seminar

DIGITAL SIGNAL PROCESSING LABORATORY USING THE ADSP-2101 MICROCOMPUTER

Vinay K. Ingle
John G. Proakis
Northeastern University

PRENTICE HALL, Englewood Cliffs, NJ 07632

Published by Prentice-Hall, Inc.
A Division of Simon & Schuster
Englewood Cliffs, New Jersey 07632

The publisher offers discounts on this book when ordered
in bulk quantities. For more information, write:

        Special Sales/College Marketing
        Prentice-Hall, Inc.
        College Technical and Reference Division
        Englewood Cliffs, New Jersey 07632

Printed in the United States of America

10 9 8 7 6 5 4 3 2 1

ISBN   0-13-218181-9   NBZI

Prentice-Hall International (UK) Limited, *London*
Prentice-Hall of Australia Pty. Limited, *Sydney*
Prentice-Hall Canada Inc., *Toronto*
Prentice-Hall Hispanoamericana, S.A., *Mexico*
Prentice-Hall of India Private Limited, *New Delhi*
Prentice-Hall of Japan, Inc., *Tokyo*
Simon & Schuster Asia Pte. Ltd., *Singapore*
Editora Prentice-Hall do Brasil, Ltda., *Rio de Janeiro*

Table of Contents

Preface

Over the past several decades, the field of digital signal processing (DSP) has grown from a theoretical infancy to a powerful practical tool and matured into an economical yet successful technology. A major reason for its success in practice is due to the development of low cost digital hardware and in particular, special purpose single chip DSP microprocessors and microcomputers. Therefore an effective education in DSP must include not only the theory but also a practical element in a laboratory environment. This book is intended to provide such an element and bridge the gap between theory and practice. The laboratory experiments and projects in this book are based on the ADSP-2101, a DSP microcomputer manufactured by Analog Devices, Inc. In addition the book is designed as a manual containing brief yet sufficient information on the total system development aspects of the ADSP-2101.

We assume that the student (or user) is familiar with the fundamentals of discrete linear systems and concurrently taking a course in DSP. These fundamentals are not covered in this book since several excellent books on DSP are available. We also assume that the student (or user) is knowledgeable in basic computer programming although no prior exposure to the actual ADSP-2101 assembly language is expected. Since the development of the experiments and projects is performed on a personal computer (PC), some knowledge of the PC and its operating system (DOS) is essential.

The following is a list of chapters and a brief description of their contents:

Chapter 1, Introduction to the ADSP-2100/2101 Family: This chapter provides a ready-reference on the building blocks of the microcomputer. Included is a brief description of the core internal architecture of the ADSP-2100 family containing computational units, data address generators, and program sequencer; a summary of the unique additional features of the ADSP-2101 including timer, serial ports, and memories; an illustration of a basic system configuration with system and memory interfacing.

Chapter 2, ADSP-2101 Instruction Set Overview: This chapter provides sufficient information to understand the nature of programming the ADSP-2101 and the capabilities of its instruction set. Included is a comprehensive summary with examples of computational instructions, data move instructions, program control instructions, multifunction instructions, and other miscellaneous instructions; a description of data structures used in programming.

Chapter 3, Overview of Development Tools: Included is a description of development tools for translating and debugging DSP source code: target system builder, the assembler, the linker, the simulator, the in-circuit emulator EZ-ICE, and the evaluation board EZ-LAB; a discussion on host computer requirements.

Chapter 4, Getting Started with the ADSP-2101: This chapter provides hands-on training on important aspects of system development. It includes learning of target system description and specification; management of simulator window environment and navigation; a description of simulator commands; instruction set work-out; a complete description of EZ-ICE firmware and window commands.

Chapter 5, Laboratory Experiments Using the ADSP-2101: This chapter is devoted to experiments which incorporate all basic operations done in DSP. It deals with simple programs and experiments in A/D and D/A conversion, signal delay and echo generation, convolution operation and recursive filtering, and waveform generation.

Chapter 6, FIR Filter Implementations: This chapter contains projects on FIR filters. Included is an overview of finite impulse response filter structures and design techniques; a single-precision direct-form implementation; a double-precision implementation; a lattice filter implementation; a single sideband modulator.

Chapter 7, IIR Filter Implementations: This chapter contains projects on IIR filters. Included is a summary of infinite impulse response filter structures and design methods; a direct form implementation; a cascade form implementation; an all-pole lattice filter implementation.

Chapter 8, Fast Fourier Transform Implementations: Included is a review of the discrete Fourier transform (DFT); a complete description and implementation of decimation-in-time and decimation-in-frequency fast Fourier transform algorithms; the inverse DFT and its implementation.

Chapter 9, Applications in Communications: This chapter focuses on several experiments dealing with waveform representation and coding, and with digital communications. Included is a description of pulse code modulation (PCM), differential PCM (DPCM) and adaptive DPCM (ADPCM), delta modulation (DM) and adaptive DM (ADM), linear predictive coding (LPC); generation and detection of dual-tone multifrequency (DTMF) signals; a description of signal detection applications in binary communications and spread spectrum communications.

Chapter 10, Adaptive Filters and their Applications: This chapter provides a formulation of experiments in the applications of adaptive filtering. Included is an introduction to the theory and implementation of adaptive FIR filters with applications to system identification, interference suppression, narrowband frequency enhancement, adaptive equalization and echo cancellation.

The book is an outgrowth of our teaching of an undergraduate laboratory course in DSP. This laboratory course containing eleven $3\frac{1}{2}$ hour sessions is taken by students concurrently with the DSP course at Northeastern University. The material described in the first four chapters is covered typically in 2 to 3 sessions in a tutorial setting. The experiments described in Chapter 5 are then done in about 6 sessions. The remaining sessions are devoted to one project which a student chooses based on the material given in the remaining five chapter.

The book can also be used in a graduate DSP course with projects on FIR and IIR filters, fast Fourier transforms, and adaptive filtering. Similarly a course on communication systems can benefit from the projects described in Chapter 9. The book can also be used as a self-study guide by anyone interested in practical DSP implementations or it can be used in an industrial setup for evaluation.

The book contains several program listings. These listings are either example programs, subroutines or computer files generated by the software development tools. These listings are given in their own "Listing" blocks which are serialized and referred to in the text. The example programs and subroutines have a header information containing the name of the respective computer file. These files are provided on a diskette which is available from Analog Devices, Inc. (In order to receive this diskette call DSP Division Applications Engineering at (617) 461-3672.) In addition, several exercise solutions are also available on the diskette. The user should work on his or her exercise then compare solutions with those given on the diskette. The complete information about the files contained on the diskette is available in README.TXT file.

The software development tools described in this book are available for the IBM-AT compatible personal computers. The Cross-Software used is version 2 or later and works on any 286- or 386-based systems. The installation of the Cross-Software requires PC-DOS 3.0 or later, 640 KB memory, and the directive "FILES=25" in the CONFIG.SYS file. Additionally, a hard disk and a color display system is highly recommended.

We would like to thank Analog Devices, Inc. for its generous support of our DSP Laboratory and encouragement for converting our lab manual into this book. The staff of the DSP Division and its Applications Group at Analog Devices, Inc. provided some material and programs reported in this book. In particular we would like to thank Bob Fine and Steve Cox who patiently answered our many questions, gave advice and explanations, and provided feedback on the earlier versions of the manuscript.

We are also indebted to our graduate students Eric Seto, Anil Shrestha, and Yiduk Kwon who helped us design and execute the experiments and projects on the development system. Finally, we wish to thank Mr. Hans Rempel of Analog Devices, Inc. who helped on the final preparation of the manuscript.

Vinay K. Ingle
John G. Proakis

Boston, Massachusetts

chapter 1

INTRODUCTION TO THE
ADSP-2100/2101 FAMILY

1.1 INTRODUCTION

The ADSP-21xx is a family of programmable single-chip processors optimized for digital signal processing (DSP) and other high-speed numeric processing applications. These processors use a modified Harvard architecture with separate buses for data and instructions. They also incorporate computational units, data address generators and a program sequencer in one device along with some necessary electronics.

The ADSP-2100 is a single-chip *microprocessor* with both program and data buses extending off chip. It requires external data and program memories, and contains three full function and independent computational units: an arithmetic/logic unit, a multiplier/accumulator and a barrel shifter. These computational units process 16-bit data directly and provide for multiprecision computation. Two data address generators and program sequencer provide address and together they allow computational operations to execute with maximum efficiency. Figure 1-1 shows the ADSP-2100 internal architecture.

The ADSP-2101 on the other hand is a single-chip *microcomputer* based on the ADSP-2100 and contains additional on-chip program and data memory, two serial ports, a timer and extensive interrupt capabilities. It has 1K words of (16-bit) data memory RAM and 2K words of (24-bit) program memory RAM on the chip. The processor can fetch an operand from on-chip data memory, an operand from on-chip program memory and the next instruction from the on-chip program memory in one single cycle. This internal bus structure is extended off-chip via a single external memory address bus and data bus. Figure 1-2 is an overall block diagram of the ADSP-2101.

The ADSP-2101 microcomputer is fabricated in a high-speed 1.0 micron double-layer metal CMOS process and operates at internal clock rate of 50MHz. With an external clock on crystal at 12.5 MHz, every instruction executes in a single cycle of 80 ns. Fabrication in CMOS results in low power requirements. The ADSP-2101 dissipates less than 1W under all conditions and no more than 80mW under standby conditions. It is available in a 68-pin Pin Grid Array (PGA) and a 68-lead Plasic Leaded Chip Carrier (PLCC).

1

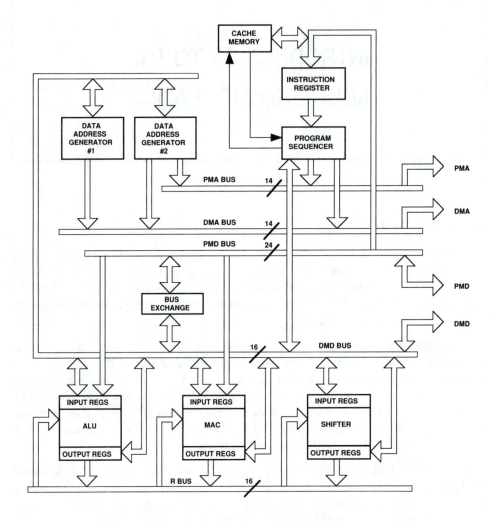

Figure 1-1: ADSP-2100 Internal Architecture

In this book, we primarily deal with the ADSP-2101 since its architecture is a superset of the ADSP-2100. When needed we will discuss the differences between the two devices. However the discussion will be based exclusively on the ADSP-2101. In the following section we provide a brief description of the internal architecture of the ADSP-2101. For a detailed description of all hardware elements, refer to *ADSP-2101/2101 Architecture User's Manual* [1]. In Section 1.3 we discuss basic system configuration with the ADSP-2101. Finally in Section 1.4, we summarize this chapter with some features of the ADSP-2101.

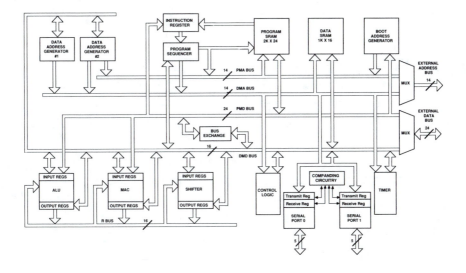

Figure 1-2: ADSP-2101 Internal Architecture

1.2 ARCHITECTURE OVERVIEW

The ADSP-2101 evolved from its predecessor, the ADSP-2100. For compatibility with the ADSP-2100, the additional features of the ADSP-2101 appear in the form of new mode controls, new processor registers and a group of memory mapped control registers. Hence we first describe the core internal architecture which is common to both processors. We then describe additional hardware elements which are special to the ADSP-2101.

1.2.1 Core Internal Architecture

Both the ADSP-2100 and the ADSP-2101 share a basic set of hardware elements called the core architecture. These elements are:

- Arithmetic-Logic Unit (ALU)
- Multiplier-Accumulator (MAC)
- Barrel Shifter
- Two Data Address Generators (DAG)
- Program Sequencer.

Efficient data transfer is achieved with the use of five internal buses:

- Program Memory Address (PMA) Bus
- Program Memory Data (PMD) Bus
- Data Memory Address (DMA) Bus
- Data Memory Data (DMD) Bus
- Result (R) Bus.

Figure 1-3 illustrates the block diagram of this core internal architecture.

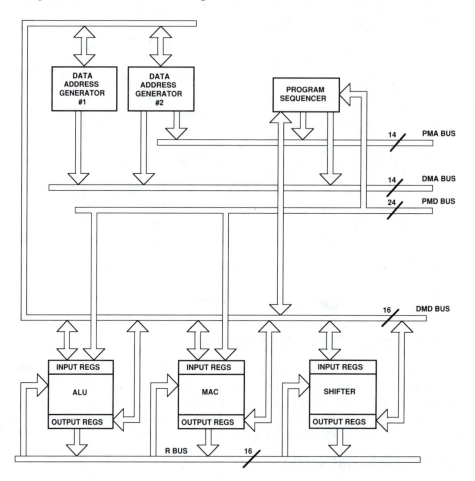

Figure 1-3: ADSP-2100 Family Core Internal Architecture

The ALU performs a standard set of arithmetic and logic operations in addition to division primitives. The MAC performs single-cycle multiply, multiply/add and multiply/subtract operations. The Shifter performs logical and arithmetic shifts, normalization, denormalization, and derive exponent operations. The Shifter implements numeric format control including multiword floating point representations. The computational units are arranged side-by-side instead of serially so that the output of any unit may be the input of any unit on the next cycle. The internal result (R) bus directly connects the computational units to make this possible.

All three sections contain input and output registers which are accessible from the internal Data Memory Data (DMD) bus. Computational operations generally take their operands from input registers and load the result into an output register. The registers act as a stopover point for data between memory and the computational circuitry. This feature introduces one level of pipelining on input, and one level on output. The R bus allows the result of a previous computation to be used directly as the input to another computation. This avoids excessive pipeline delays when a series of different operations are performed.

Two dedicated data address generators and a powerful program sequencer ensure efficient use of these computational units. The Data Address Generators (DAGs) provide memory addresses when memory data is transferred to or from the input/output registers. Each DAG keeps track of up to four address pointers. When a pointer is used for indirect addressing, it is post-modified by a value in a specified register. With two independent DAGs, the processor can generate two addresses simultaneously for dual operand fetches.

A length value may be associated with each pointer to implement automatic modulo addressing for circular buffers. (The circular buffer feature is also used by the serial ports for automatic data transfers. Refer to Section 1.2.2 for information on Serial Ports.) DAG1 can supply addresses to data memory only. DAG2 can supply addresses to either the data memory or the program memory. Two independent address generators allow for simultaneous access of data stored in the program memory and data stored in the data memory.

The Program sequencer supplies instruction addresses to the program memory. The sequencer is driven by the Instruction Register which holds the currently executing instruction. The instruction register introduces a single level of pipelining into the program flow. Instructions are fetched,loaded into the instruction register, and decoded during one processor cycle; and executed during the following cycle while the next instruction is prefetched. To minimize overhead cycles, the sequencer supports conditional jumps, subroutine calls and returns in a single-cycle. With an internal loop counter and loop stack, the ADSP-2101 executes looped code with zero-overhead. No explicit jump instructions are required to loop.

These components are supported by five internal buses: The PMA and DMA buses are used internally for the addresses associated with Program and Data Memory. The Program Memory Data (PMD) and Data Memory Data (DMD) buses are used for the data associated with the memory spaces. These two pairs of buses are multiplexed off chip to the external address and data buses. The $\overline{BMS}$, $\overline{DMS}$ and $\overline{PMS}$ (pin) signals select the different address spaces. The R bus is an internal bus which serves to transfer intermediate results directly between the various computational sections.

The Program Memory Address (PMA) bus is 14 bits wide allowing direct access of up to 16K words of mixed instruction code and data. The program memory data (PMD) is 24 bits wide to accommodate the 24-bit instruction width.

The Data Memory Address (DMA) bus is 14 bits wide allowing direct access of up to 16 K words of data. The Data Memory Data (DMD) bus is 16 bits wide. The data memory data (DMD) bus provides a path for the contents of any register in the processor to be transferred to any other register or to any external data memory location in a single cycle. The data memory address comes from two sources: an absolute value specified in the instruction code (direct addressing) or the output of a data address generator (indirect addressing). Only indirect addressing is supported for data fetches from program memory.

The Program Memory data (PMD) bus can also be used to transfer data to and from the computational units through direct paths or via the PMD-DMD bus exchange unit. The PMD-DMD bus exchange unit permits data to be passed from one bus to the other. It contains hardware (PX register) to overcome the 8-bit width discrepancy between the two buses, if necessary. The 8-bit PX register can be read or written as any other data register.

Arithmetic-Logic Unit (ALU)

The Arithmetic/Logic Unit (ALU) provides a standard set of arithmetic and logical functions. The arithmetic functions are add, subtract, negate, increment, decrement and absolute value. These are supplemented by two division primitives with which multiple cycle division can be constructed. The logic functions are AND, OR, XOR (exclusive OR) and NOT. Figure 1-4 shows a block diagram of the ALU.

The ALU is 16 bits wide with two 16-bit input ports, X and Y, and one output port, R. The ALU accepts a carry-in signal (CI) which is the carry bit from the processor arithmetic status register (ASTAT). The ALU generates six status signals: the zero (AZ) status, the negative (AN) status, the carry (AC) status, the overflow (AV) status, the X-input sign (AS) status, and the quotient (AQ) status. All arithmetic status signals are latched into the arithmetic status register (ASTAT) at the end of the cycle.

The X input port of the ALU can accept data from two sources: the AX register file or the result (R) bus. The R bus connects the output registers of all the computational units, permitting them to be used as input operands directly. The AX register file is dedicated to the X input port and consists of two registers, AX0 and AX1. These AX registers are readable and writable from the DMD bus. The instruction set also provides for reading these registers over the PMD bus, but there is no direct connection; this operation uses the DMD-PMD bus exchange unit. The AX register file outputs are dual-ported so that one register can provide input to the ALU while either one simultaneously drives the DMD bus.

The Y input port of the ALU can also accept data from two sources: the AY register file and the ALU feedback (AF) register. The AY register file is dedicated to the Y input port and consists of two registers, AY0 and AY1. These registers are readable and writable from the DMD bus and writable from the PMD bus. The instruction set also provides for reading these registers over the PMD bus, but there is no direct connection; this operation uses the DMD-PMD bus exchange unit. The AY register file outputs are also dual-ported: one AY register can provide input to the ALU while either one simultaneously drives the DMD bus.

The output of the ALU is loaded into either the ALU feedback (AF) register or the ALU result (AR) register. The AF register is an ALU internal register which allows the ALU result to be used directly as the ALU Y input. The AR register can drive both the DMD bus and the R bus. It is also loadable directly from the DMD bus. The instruction set also provides for reading AR over the PMD bus, but there is no direct connection; this operation uses the DMD-PMD bus exchange unit.

Any of the registers associated with the ALU can be both read and written in the same cycle. Registers are read at the beginning of the cycle and written at the end of the cycle. A register read, therefore, reads the value loaded at the end of a previous cycle. A new value written to a register cannot be read out until a subsequent cycle. This allows an input register

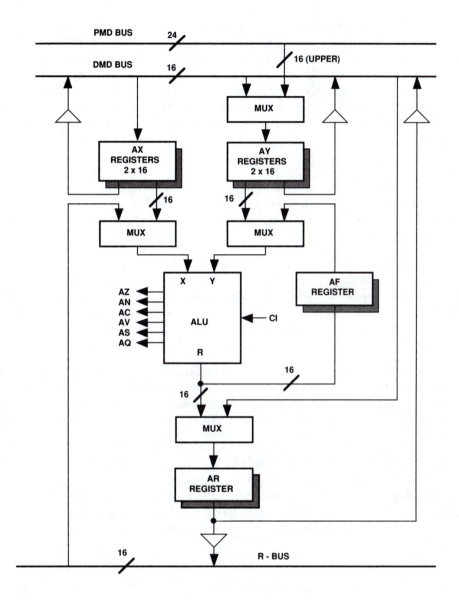

Figure 1-4: ALU Block Diagram

to provide an operand to the ALU at the beginning of the cycle and be updated with the next operand from memory at the end of the same cycle. It also allows a result register to be stored in memory and updated with a new result in the same cycle. See the discussion of "Multifunction Instructions" in the Chapter 2, "ADSP-2101 Instruction Set Overview", for an illustration of this same-cycle read and write.

The ALU contains a duplicate bank of registers, shown in Figure 1-4 behind the primary registers. There are actually two sets of AR, AF, AX, and AY register files. Only one bank is accessible at a time. The additional bank of registers can be activated (such as during an interrupt service routine) for extremely fast context switching. A new task, like an interrupt service routine, can be executed without transferring current states to storage. The selection of the primary or alternate bank of registers is controlled by bit 0 in the processor mode status register (MSTAT). If this bit is a 0, the primary bank is selected; if it is a 1, the secondary bank is selected.

Multiplier-Accumulator (MAC)

The Multiplier/Accumulator provides high-speed multiplication, multiplication with cumulative addition, multiplication with cumulative subtraction, saturation and clear-to-zero functions. A feedback function allows part of the accumulator output to be directly used as one of the multiplicands on the next cycle. Figure 2.5 shows a block diagram of the multiplier/accumulator.

The multiplier has two 16-bit input ports X and Y, and a 32-bit product output port P. The 32-bit product is passed to a 40-bit adder/subtractor which adds or subtracts the new product from the content of the multiplier result (MR) register, or passes the new product directly to MR. The MR register is 40-bits wide. In this manual, we refer to the entire register as MR. The register actually consists of three smaller registers: MR0 and MR1, which are 16 bits wide, and MR2, which is 8 bits wide.

The adder/subtractor is greater than 32 bits to allow for intermediate overflow in a series of multiply/accumulate operations. The multiply overflow (MV) status bit is set when the accumulator has overflowed beyond the 32-bit boundary, that is, when there are significant (non-sign) bits in the top nine bits of the MR register (based on twos-complement arithmetic).

The input/output registers of the MAC section are similar to the ALU. The X input port can accept data from either the MX register file or from any register on the result (R) bus. The R bus connects the output registers of all the computational units, permitting them to be used as input operands directly. There are two registers in the MX register file, MX0 and MX1. These registers can be read and written from the DMD bus. The MX register file outputs are dual-ported so that one register can provide input to the multiplier while either one simultaneously drives the DMD bus.

The Y input port can accept data from either the MY register file or the MF register. The MY register file has two registers, MY0 and MY1; these registers can be read and written from the DMD bus and written from the PMD bus. The ADSP-2101 instruction set also provides for reading these registers over the PMD bus, but there is no direct connection; this operation uses the DMD-PMD bus exchange unit. The MY register file outputs are also dual-ported so that one register can provide input to the multiplier while either one simultaneously drives the DMD bus.

The output of the adder/subtractor goes to either the MF register or the MR register. The MF register is a feedback register which allows bits 16-31 of the result to be used directly as

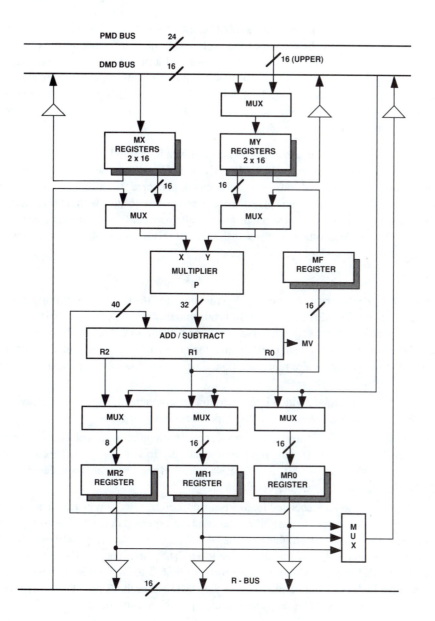

Figure 1-5: MAC Block Diagram

the multiplier Y input on a subsequent cycle. The 40-bit adder/subtractor register (MR) is divided into three sections: MR2, MR1, and MR0. Each of these registers can be loaded directly from the DMD bus and output to either the DMD bus or the R bus.

Any of the registers associated with the MAC can be both read and written in the same cycle. Registers are read at the beginning of the cycle and written at the end of the cycle. A register read, therefore, reads the value loaded at the end of a previous cycle. A new value written to a register cannot be read out until a subsequent cycle. This allows an input register to provide an operand to the MAC at the beginning of the cycle and be updated with the next operand from memory at the end of the same cycle. It also allows a result register to be stored in memory and updated with a new result in the same cycle.

The MAC contains a duplicate bank of registers, shown in Figure 1-5 behind the primary registers. There are actually two sets of MR, MF, MX, and MY register files. Only one bank is accessible at a time. The additional bank of registers can be activated for extremely fast context switching. A new task, such as an interrupt service routine, can be executed without transferring current states to storage. The selection of the primary or alternate bank of registers is controlled by bit 0 in the processor mode status register (MSTAT). If this bit is a 0, the primary bank is selected; if it is a 1, the secondary bank is selected.

Barrel Shifter

The shifter unit provides a complete set of shifting functions for 16-bit inputs, yielding a 32-bit output. These include arithmetic shift, logical shift and normalization. The Shifter also performs derivation of exponent and derivation of common exponent for an entire block of numbers. These basic functions can be combined to efficiently implement any degree of numerical format control, including full floating-point representation. Figure 1-6 shows a block diagram of the shifter.

The shifter section can be divided into the following components: the shifter array, the OR/PASS logic, the exponent detector, and the exponent compare logic.

The shifter array is a 16x32 barrel shifter. It accepts a 16-bit input and can place it anywhere in the 32-bit output field, from off-scale right to off-scale left, in a single cycle. This gives 49 possible placements within the 32-bit field. The placement of the 16 input bits is determined by a control code (C) and a HI/LO reference signal.

The shifter array and its associated logic are surrounded by a set of registers. The shifter input (SI) register provides input to the shifter array and the exponent detector. The SI register is 16 bits wide and is readable and writable from the DMD bus. The shifter array and the exponent detector also takes as inputs AR, SR or MR via the R bus. The shifter result (SR) register is 32 bits wide and is divided into two 16-bit sections, SR0 and SR1. The SR0 and SR1 registers can be loaded from the DMD bus and output to either the DMD bus or the R bus. The SR register is also fed back to the OR/PASS logic to allow double-precision shift operations.

The SE register ("shifter exponent") is 8 bits wide and holds the exponent during the normalize and denormalize operations. The SE register is loadable and readable from the lower 8 bits of the DMD bus. It is a twos-complement, integer value.

The SB register ("shifter block") is important in block floating-point operations where it holds the block exponent value, that is, the value by which the block values must be shifted to normalize the largest value. SB is 5 bits wide and holds the most recent block exponent value. The SB register is loadable and readable from the lower 5 bits of the DMD bus. It is a twos-complement, integer value.

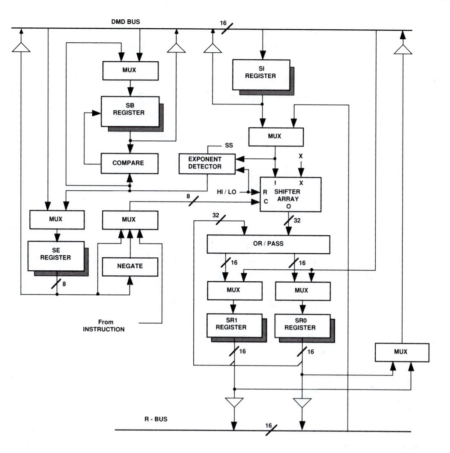

Figure 1-6: Shifter Block Diagram

Whenever the SE or SB registers are output onto the DMD bus, they are sign-extended to form a 16-bit value. Any of the SI, SE or SR registers can be read and written in the same cycle. Registers are read at the beginning of the cycle and written at the end of the cycle. All register reads, therefore, read values loaded at the end of a previous cycle. A new value written to a register cannot be read out until a subsequent cycle. This allows an input register to provide an operand to the Shifter at the beginning of the cycle and be updated with the next operand at the end of the same cycle. It also allows a result register to be stored in memory and updated with a new result in the same cycle.

The shifter section contains a duplicate bank of registers, shown in Figure 1-6 behind the primary registers. There are actually two sets of SE, SB, SI, SR1, and SR0 registers. Only one bank is accessible at a time. The additional bank of registers can be activated for extremely fast context switching. A new task, such as an interrupt service routine, can then be executed without transferring current states to storage. The selection of the primary or alternate bank of registers is controlled by bit 0 in the processor mode status register (MSTAT). If this bit is a 0, the primary bank is selected; if it is a 1, the secondary bank is selected.

The shifting of the input is determined by a control code (C) and a HI/LO reference signal. The control code is an 8-bit signed value which indicates the direction and number of places the input is to be shifted. Positive codes indicate a left shift (upshift) and negative codes indicate a right shift (downshift). The control code can come from three sources: the content of the shifter exponent (SE) register, the negated content of the SE register or an immediate value from the instruction.

The HI/LO signal determines the reference point for the shifting. In the HI state, all shifts are referenced to SR1 (the upper half of the output field), and in the LO state, all shifts are referenced to SR0 (the lower half). The HI/LO reference feature is useful when shifting 32-bit values since it allows both halves of the number to be shifted with the same control code. HI/LO reference signal is selectable each time the shifter is used.

The shifter fills any bits to the right of the input value in the output field with zeros, and bits to the left are filled with the extension bit (X). The extension bit can be fed by three possible sources depending on the instruction being performed. The three sources are the MSB of the input, the AC bit from the arithmetic status register (ASTAT) or a zero.

The OR/PASS logic allows the shifted sections of a multiprecision number to be combined into a single quantity. When PASS is selected, the shifter array output is passed through and loaded into the shifter result (SR) register unmodified. When OR is selected, the shifter array is bitwise ORed with the current contents of the SR register before being loaded there.

The exponent detector derives an exponent for the shifter input value. The exponent detector operates in one of three ways which determine how the input value is interpreted. In the HI state, the input is interpreted as a single precision number or the upper half of a double precision number. The exponent detector determines the number of leading sign bits and produces a code which indicates how many places the input must be up-shifted to eliminate all but one of the sign bits. The code is negative so that it can become the effective exponent for the mantissa formed by removing the redundant sign bits.

In the HI-extend state (HIX), the input is interpreted as the result of an add or subtract performed in the ALU section which may have overflowed. Therefore the exponent detector takes the arithmetic overflow (AV) status into consideration. If AV is set, then a +1 exponent is output to indicate an extra bit is needed in the normalized mantissa (the ALU Carry bit); if AV is not set, then HI-extend functions exactly like the HI state. When performing a derive exponent function in HI or HI-extend modes, the exponent detector also outputs a shifter sign (SS) bit which is loaded into the arithmetic status register (ASTAT). The sign bit is the same as the MSB of the shifter input except when AV is set; when AV is set in the HI-extend state, the MSB is inverted to restore the sign bit of the overflowed value.

In the LO state, the input is interpreted as the lower half of a double precision number. In the LO state, the exponent detector interprets the SS bit in the arithmetic status register (ASTAT) as the sign bit of the number. The SE register is loaded with the output of the exponent detector only if SE contains P15. This occurs only when the upper half—which must be processed first—contained all sign bits. The exponent detector output is also offset by P16 to account for the fact that the input is actually the lower half of a 32-bit value.

The exponent compare logic is used to find the largest exponent value in an array of shifter input values. The exponent compare logic in conjunction with the exponent detector derives a block exponent. The comparator compares the exponent value derived by the exponent detector with the value stored in the shifter block exponent (SB) register and updates the SB register only when the derived exponent value is larger than the value in the SB register.

Data Address Generators (Dag)

The ADSP-2101 contains two independent data address generators so that both program and data memories can be accessed simultaneously. The DAGs provide indirect addressing capabilities and perform automatic address modification. For circular buffers, the DAGs can perform modulo address modification. The two DAGs differ: DAG1 only generates data memory addresses, but provides an optional bit-reversal capability; DAG2 can generate both data memory and program memory addresses, but has no bit-reversal capability.

Figure 1-7 shows a block diagram of a single data address generator. There are three register files: the modify (M) register file, the index (I) register file, and the length (L) register file. Each of the register files contains four 14-bit registers which can be read from and written to via the DMD bus.

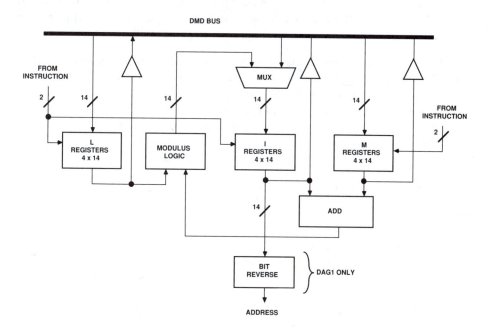

Figure 1-7: Data Address Generator Block Diagram

The I registers (I0-3 in DAG1, I4-7 in DAG2) contain the actual addresses used to access memory. When data is accessed the in indirect mode, the address stored in the selected I register becomes the memory address. With DAG1, the output address can be bit-reversed by setting the appropriate mode bit in the mode status register (MSTAT) as discussed below. Bit-reversal facilitates FFT addressing.

The data address generator employs a post-modify scheme. After an indirect data access, the specified M register (M0-3 in DAG1, M4-7 in DAG2) is added to the specified I register to generate the new I value. The choice of the I and M registers are independent within each DAG. In other words, any register in the I0-3 set may be modified by any register in the M0-3 set in any combination, but not by those in DAG2 (M4-7). The modification values stored in M registers are signed numbers so that the next address can be either higher or lower. The address generators support both linear addressing and circular addressing. The value of the L register determines which addressing scheme is used. For circular buffer addressing, the L register is initialized with the length of the buffer. For linear addressing, the modulus logic is disabled by setting the corresponding L register to zero.

L registers and I registers are paired and the selection of the L register (L0-3 in DAG1, L4-7 in DAG2) is determined by the I register used. Each time an I register is selected, the corresponding L register provides the modulus logic with the length information. If the sum of the M register content and the I register content crosses the buffer boundary, the modified I register value is calculated by the modulus logic using the L register value.

All data address generator registers (I, M, and L registers) are loadable and readable from the lower 14 bits of the DMD bus. Since I and L register contents are considered to be unsigned, the upper 2 bits of the DMD bus are padded with zeros when reading them. M register contents are signed; when reading an M register, the upper 2 bits of the DMD bus are sign-extended.

The modulus logic implements automatic pointer wraparound for accessing circular buffers. To calculate the next address, the modulus logic uses the following information.
- The current location; found in the I register (unsigned)
- The modify value; found in the M register (signed)
- The buffer length; found in the L register (unsigned)
- The buffer base address

From these inputs, the next address is calculated with the formula:

Next address = (I + M - B) Modulo (L) + B

where:

I = current address,
M = modify value (signed)
B = base address (generated by the linker)
L = buffer length M+
I = modified address
$|M| < L$ (this insures that the next address cannot wrap around the buffer more than once in one operation).

Program Sequencer

The program sequencer generates a stream of instruction addresses, and provides flexible control of program flow. It provides for zero-overhead looping, single-cycle branching (both conditional and unconditional) and sophisticated interrupt processing. Figure 1-8 shows a block diagram for the program sequencer and status sections of the ADSP-2101.

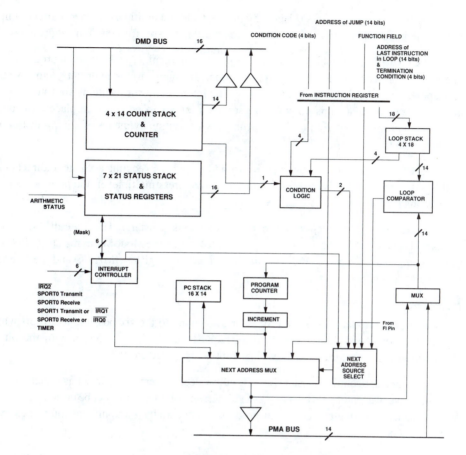

Figure 1-8: ADSP-2101 Program Sequencer

The sequencing logic controls the flow of ADSP-2101 program execution by outputting a program memory address onto the PMA bus from one of the following four possible sources: PC incrementer, PC stack, instruction register, interrupt controller. The next address source selector in the diagram controls which of these four sources are output from the next address multiplexer, based on outputs from the instruction register, condition logic, loop comparator, and interrupt controller. A fifth possibility for the next program memory address, although not part of the program sequencer, is DAG2 when a register indirect jump is executed.

The PC incrementer is selected as the source of the next program memory address if program flow is sequential. This is also the case when a conditional jump or return is not taken and when a DO UNTIL loop terminates (see below for a description of the DO UNTIL construct).

The PC stack is used as the source for the next program memory address when a return from subroutine or return from interrupt is executed. The top stack value is also used as the next program memory address when returning to the top of a DO UNTIL loop.

The instruction register is selected by the next address multiplexer when a direct jump is taken. The jump address field of the instruction word itself specifies the jump address.

The interrupt controller provides the next program memory address when processing an interrupt. Upon recognizing an interrupt, the processor jumps to the interrupt vector location corresponding to the active interrupt request. The interrupt vector locations are four program memory locations apart; this allows short service routines to be coded in place. For longer routines, control is transferred to the interrupt service routine by means of a jump instruction at the interrupt vector.

DAG2 sources the next program memory address when executing a register indirect jump. In this case, since DAG2 is not an input to the next address multiplexer, the program counter must be loaded from the PMA bus.

The program sequencer section contains six status registers. These are the Arithmetic Status register (ASTAT), the Stack Status register (SSTAT), the Mode Status register (MSTAT), the Interrupt Control register (ICNTL), the Interrupt Mast register (IMASK) and the Interrupt Force and Clear register (IFC).

Interrupts

The interrupt controller allows the processor to respond to the six possible interrupts with a minimum of overhead. Individual interrupt requests are logically ANDed with the bits in IMASK; the highest priority unmasked interrupt is then selected.

The interrupt control register, ICNTL, allows each interrupt to be set as either edge- or level-sensitive. Depending on bit 4 in ICNTL, interrupt routines can either be nested with higher priority interrupts taking precedence or processed sequentially with only one interrupt service active at a time.

The 12-bit interrupt force and clear register, IFC, is a write-only register that contains a force bit and a clear bit for each of the six possible interrupts.

When responding to an interrupt, the status registers ASTAT, MSTAT, IMASK are pushed onto the status stack and the PC counter is loaded with the appropriate vector address. The status stack is seven levels deep to allow interrupt nesting. The stack is automatically popped when a return from the interrupt is executed.

The vector addresses for each interrupt are fixed. In the ADSP-2101 each vector location identifies a block of four instructions. Short service routines can be executed without an additional JUMP, minimizing overhead.

IMASK

IMASK is six bits wide and allows the interrupt inputs to be individually enabled or disabled. The bits in IMASK are:

0   Timer interrupt enable
1    $\overline{\text{IRQ0}}$ or SPORT1 receive interrupt enable
2    $\overline{\text{IRQ1}}$ or SPORT1 transmit interrupt enable
3   SPORT0 receive interrupt enable
4   SPORT0 transmit interrupt enable
5    $\overline{\text{IRQ2}}$ interrupt enable

The bits are all positive sense (0=disabled, 1=enabled). IMASK is set to zero upon a processor reset so that all interrupts are disabled initially.

ICNTL

ICNTL is a 5-bit register configuring the interrupt modes of the processor. The bits in ICNTL are:

0    $\overline{\text{IRQ0}}$ sensitivity (if $\overline{\text{IRQ0}}$ is configured)
1    $\overline{\text{IRQ1}}$ sensitivity (if $\overline{\text{IRQ1}}$ is configured)
2    $\overline{\text{IRQ2}}$ sensitivity
3   zero
4   Interrupt Nesting Mode

The sensitivity bits determine whether a given interrupt input is edge- or level-sensitive (0 = level-sensitive, 1 = edge-sensitive).

The interrupt nesting mode determines whether higher priority interrupt service routines are automatically nested. When the nesting mode bit is cleared, all IMASK bits are automatically cleared when an interrupt service routine is entered, so that all interrupts are masked (no nesting can occur). The previous IMASK value is pushed on the stack. When the nesting mode bit is set, only the bits in IMASK for equal and lower priority interrupts are cleared. Higher priority interrupts that are not masked can interrupt the current interrupt service routine. This value of IMASK can also be changed at any time to allow other nesting schemes.

Edge-triggered interrupts are automatically cleared when the interrupt service routine is called. They can also be cleared by writing a one to the appropriate IFC bit. The Timer and Serial Port interrupts act as edge-sensitive interrupts which can be masked, cleared or forced with software.

IFC

The write-only IFC register is twelve bits wide and contains a bit for clearing and a bit for forcing each of the six possible interrupts in the ADSP-2101. The bits in IFC are defined as follows.

Bit 0   Timer interrupt clear
Bit 1   SPORT1 receive or $\overline{\text{IRQ0}}$ interrupt clear
Bit 2   SPORT1 transmit or $\overline{\text{IRQ1}}$ interrupt clear
Bit 3   SPORT0 receive interrupt clear
Bit 4   SPORT0 transmit interrupt clear
Bit 5 $\overline{\text{IRQ2}}$ interrupt clear

Bit 6   Timer interrupt force
Bit 7   SPORT1 receive or $\overline{\text{IRQ0}}$ interrupt force
Bit 8   SPORT1 transmit or $\overline{\text{IRQ1}}$ interrupt force
Bit 9   SPORT0 receive interrupt force
Bit 10   SPORT0 transmit interrupt force
Bit 11 $\overline{\text{IRQ2}}$ interrupt force

Pending edge-sensitive interrupts can be cleared by writing a one to the appropriate clear bit (0-5) in IFC. Edge-triggered interrupts are cleared automatically when the corresponding interrupt service routine is called.

Edge-sensitive interrupts can be forced under program control by writing a one to the force bit (6-11) corresponding to the desired interrupt. This causes the chip to respond as though the interrupt had occurred. Forcing a level-sensitive interrupt has no effect. The Timer and SPORT interrupts behave like edge-sensitive interrupts and can be masked, cleared and forced.

Loop Mechanisms

Loop stack and comparator provides the zero-overhead looping mechanism. A DO UNTIL instruction contains the address for the end of the loop and the termination condition. When a DO UNTIL instruction is executed, this information is loaded into the loop stack, and the PC value is pushed onto the PC stack after being incremented. The loop comparator compares the end of loop address with the next address, and signals the end of loop when the two are equal. The processor then checks if the termination condition is met. Depending on this condition, the next address selector chooses between the PC stack (jump to beginning of loop) and the PC incrementer (fall out of loop). The loop stack is four level deep, permitting four levels of zero-overhead loop nesting.

The down counter and the count stack also support this powerful looping mechanism. The down counter is a 14-bit register with auto-decrement capability. It is loaded from the DMD bus with the loop count. The count is decremented every time the counter value is checked; when the count expires, the counter expired (CE) flag is set. The count stack allows the nesting of loops by storing temporarily dormant loop counts. When a new value is loaded into the counter from the DMD bus, the current counter value is automatically pushed onto the count stack, as program flow enters a loop. The count stack is automatically popped whenever the CE flag is tested and is true, thereby resuming execution of the code outside the loop. It is also possible to overwrite the counter, without pushing its value on the count stack, if loop nesting is not occurring.

Status Registers

The ADSP-2101 maintains six status registers, which can be accessed over the DMD bus (one is read-only and one is write-only, however). These registers are:

ASTAT   Arithmetic Status register
SSTAT   Stack Status register   *(read-only)*
MSTAT   Mode Status register

ICNTL   Interrupt Control register
IMASK   Interrupt Mask register
IFC   Interrupt Force and Clear   (*write-only*)

The interrupt registers are described in a previous section; the other three are discussed below.

ASTAT

ASTAT is 8 bits wide and holds the status information generated by the computational sections of the processor. The bits in ASTAT are defined as follows:

0   AZ   (ALU result zero)
1   AN   (ALU result negative)
2   AV   (ALU overflow)
3   AC   (ALU carry)
4   AS   (ALU X input sign)
5   AQ   (ALU quotient flag)
6   MV   (MAC overflow)
7   SS   (Shifter input sign)

The bits are positive sense (1=true, 0=false). They are automatically updated when a new status is generated by the arithmetic operations affecting them, as defined by the following table:

Status Bit	Updated on:
AZ, AN, AV, AC	Any ALU operation except division
AS	ALU absolute value operation
AQ	ALU divide operations
MV	Any MAC operation except saturate MR
SS	Shifter exponent detect operation

The computation condition codes are described in Chapter 2.

SSTAT

SSTAT is 8 bits wide and holds the status of the four internal stacks. The bits in SSTAT are:

0   PC Stack Empty
1   PC Stack Overflow
2   Count Stack Empty
3   Count Stack Overflow
4   Status Stack Empty
5   Status Stack Overflow
6   Loop Stack Empty
7   Loop Stack Overflow

All of the bits are positive sense (1=true, 0=false). The empty status bits indicate that the stack is empty. The overflow status bits indicate that the stack has overflowed. Since the stack overflow status bits "stick" once they are set, subsequent pop operations have no effect on them. This means that the stack can be both overflowed and empty under certain circumstances. A processor reset or a software reboot must be executed to clear the stack overflow status.

MSTAT

MSTAT is a 7-bit register that defines various operating modes of the processor. The Mode Control instruction enables or disables the operating modes. The bits in MSTAT are:

0	Data Register Bank Select
1	Bit Reverse Mode (DAG1 only)
2	ALU Overflow Latch Mode
3	AR Saturation Mode
4	MAC Result P Placement Mode
5	Timer Enable
6	Go Mode

The data register bank select bit determines which set of data registers is currently active (0=primary, 1=secondary). The data registers include all of the result and input registers to the ALU, MAC, and Shifter (AX0, AX1, AY0, AY1, AF, AR, MX0, MX1, MY0, MY1, MF, MR0, MR1, MR2, SB, SE, SI, SR0 and SR1). At $\overline{RESET}$, the data register bank select bit is cleared. The bit reverse mode, when enabled, bit-wise reverses all addresses generated by DAG1. This is most useful for re-ordering the input or output data in a radix-2 FFT algorithm. The ALU overflow latch mode causes the AV (ALU overflow) status bit to "stick" once it is set. In this mode, when an ALU overflow occurs, AV will be set and remain set, even if subsequent ALU operations do not generate overflows. AV can then only be cleared by writing a zero into it from the DMD bus. The AR saturation mode, when set, causes ALU results to be saturated to the maximum positive (0x7FFF[1]) or negative (0x8000) values when an ALU overflow or underflow occurs. The MAC Result P Placement bit, when set to 0, results in the ADSP-2100 result placement of the multiplier product in the MR register (one bit shift). When this bit is 1, no shift occurs. The Timer Enable bit, when set to 1, enables the timer decrement mechanism. The Go Mode bit, when set to 1, allows the processor to continue operations internally (when possible) while the external address and data buses are tristated during a bus grant.

1.2.2 Additional Features of the ADSP-2101

Being a microcomputer, the ADSP-2101 contains supplementary internal hardware elements so that a basic system configuration can be built with a minimum number of external devices. These elements include a timer, two serial ports (SPORT), boot address generator, program and data memories.

Timer

The ADSP-2101 contains a programmable interval timer to generate periodic interrupts based on multiples of the processor's cycle time. Figure 1-9 shows the timer block diagram. It includes

---

1 Throughout this book, hexadecimal numbers are denoted by "0x" prefix.

two 16-bit registers, TCOUNT and TPERIOD and one 8-bit register TSCALE. These registers are memory mapped. The extended mode control instruction enables and disables the timer by setting and clearing bit 5 in the mode status register, MSTAT.

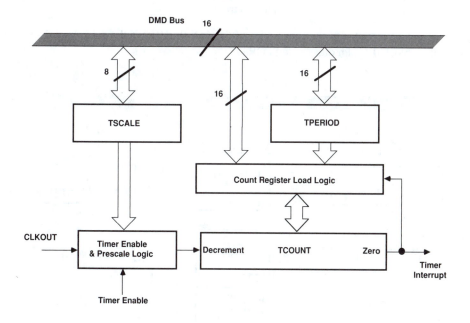

Figure 1-9: Timer Block Diagram

TCOUNT is a count register. When the timer is enabled, it is decremented as often as once every instruction cycle. When the counter reaches zero, an interrupt is generated. TCOUNT is then reloaded from the TPERIOD register and the count begins again. TSCALE stores a scaling value that is one less than the number of cycles between decrements of TCOUNT. For example, if the value in TSCALE register is 0, the counter register decrements once every cycle. If the value in TSCALE is 1, the counter decrements once every 2 cycles. In a processor with an 80ns cycle time, for example, the timer interrupt could occur as infrequently as every 1.34 seconds if a maximum scaling value is used. With an 80ns resolution, a maximum period of 5.24ms can be timed.

Serial Ports

The ADSP-2101 incorporates two complete serial ports, SPORT0 and SPORT1, for serial communications and multiprocessor coordination. Each serial port has a 5-pin interface consisting of the following signals.

Signal Name	Function
SCLK	Serial Clock I/O
RFS	Receive frame synch I/O
TFS	Transmit frame synch I/O
DR	Serial data receive
DT	Serial data transmit

Here is a brief list of the capabilities of the ADSP-2101 SPORTs. Figure 1-10 shows a simplified block diagram of a single SPORT.

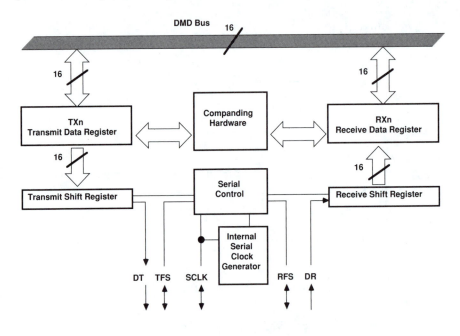

Figure 1-10: Serial Port Block Diagram

- Bidirectional: each SPORT has a separate transmit and receive section.

- Double-buffered: each SPORT section (both receive and transmit) has a data register accessible to the user and an internal transfer register. The double-buffering provides additional time to service the SPORT.

- Flexible clocking: each SPORT can use an external serial clock (from 0 Hz to the processor frequency) or generate its own (from $1/2^{17}$ of the processor frequency to 1/2 the processor frequency).

- Flexible framing: Framings for the receive and transmit sections on each SPORT are independent. Each section can run in a frameless mode, with internally-generated or externally-generated frame synch signals, with active high or inverted frame signals, and with either of two pulse widths or timings. The receive and transmit sections share the same serial clock.

- Flexible word length: each SPORT supports serial data word lengths from three to sixteen bits.

- Companding in hardware: each SPORT provides optional A-law and μ-law companding. Different companding can be used for each SPORT, for example, A-law for SPORT0 and μ-law for SPORT1.

- Flexible interrupt scheme: each SPORT section (receive and transmit) can generate a unique interrupt upon completing a data word transfer or after transferring an entire buffer (see next item).

- Auto-buffering with single-cycle overhead: using the ADSP-2101 DAGs, each SPORT can receive and/or transmit an entire circular buffer of data with an overhead of only one cycle per data word. Transfers to and from the SPORT and the circular buffer are automatic in this mode and do not require additional programming. An interrupt is generated only when pointer wraparound occurs in the circular buffer.

- Multichannel capability: SPORT0 provides a multichannel interface for selective receipt and transmission of arbitrary data channels from a twenty-four or thirty-two word, time-division multiplexed, serial bitstream. This is especially useful for T1 or CEPT interfaces or as a network communication scheme for multiple processors.

- Alternate configuration: SPORT1 can be configured as two external interrupt inputs ($\overline{\text{IRQ0}}$ and $\overline{\text{IRQ1}}$) and the Flag In and Flag Out signals. The internally generated serial clock may still be used in this configuration.

Each SPORT has a receive and a transmit register; SPORT0's registers are RX0 and TX0, SPORT1's are RX1 and TX1. Companding (a contraction of COMpressing and exPANDing) is the process of logarithmically encoding data to reduce the number of bits that must be sent. Both SPORTs share the companding hardware: one expansion and one compression operation can occur in each processor cycle. In the event of contention, SPORT0 has priority. The ADSP-2101 supports both of the widely used algorithms for companding: A-law and μ-law. The type of companding can be independently selected for each SPORT.

The TXn and RXn registers are identified by name in the ADSP-2101 assembly language, not memory-mapped. TXn and RXn can be read and written (like other non-data registers) with the following instruction types: read/write to data memory (direct address), load non-data immediate, and internal (register-to-register) moves.

There are two ways to generate the SPORT interrupts: after the transmit or receipt of 1) each word (normal word by word operation) or 2) each complete buffer of data words (autobuffer operation).

These serial port features, in conjunction with other features of the ADSP-2101, make it possible to interface to most codecs, A/Ds, DACs and to additional ADSP-2101s with no additional hardware and little or no software overhead.

Memories

The ADSP-2101 has three separate memory spaces: data memory, program memory and boot memory. Boot memory is only active during the loading of program code from an external device (ROM, EPROM, or RAM). Program memory consists of a single address space 24-bit (3-byte) wide, 2K of which resides on the chip and the rest is external. This program memory is dual purpose for both instruction and data storage. Hence two program memory locations can be accessed in a single cycle. Data memory is a single address space 16-bit (2-byte) wide, 1K of which resides on the chip and the rest is external. There are separate program and data buses on the chip.

Boot Address Generator

To execute the boot operation, the boot address generator generates the appropriate byte addresses and loads the ADSP-2101 internal program memory with the contents of the external EPROM. The ADSP-2101 internal program memory is loaded beginning with the high addresses. Although 2K words 3-byte wide internal program memory requires only 6K bytes of storage, boot memory is organized into eight pages which are 8K bytes long. Every fourth byte of a page is an "empty" byte, except the first one, which contains the page length. The page length is read first and then bytes are loaded from the top of the page downwards. This results in a shorter booting times for shorter pages. The boot address generator is designed to generate the proper sequence of addresses.

PMD-DMD Bus Exchange

This unit couples the program memory data bus and the data memory data bus, allowing them to transfer data in both directions. Since the program memory data (PMD) bus is 24 bits wide, while the data memory data (DMD) bus is 16 bits wide, only the upper 16 bits of PMD can be directly transferred. An internal register (PX) is loaded with (or supplies) the additional 8 bits. This register can be directly loaded or read when the full 24 bits are required.

Figure 1-11 shows a block diagram of this circuit. There are two types of connections provided in this circuit.

The first type of connection is a one-way path from each bus to the other. This is implemented with two tristate buffers connecting the DMD bus with the upper 16 bits of the PMD bus. One of these two buffers is normally used when data is exchanged between the program memory and one of the registers connected to the DMD bus. This is the path used to write data to program memory; it is not shown in the individual computational unit block diagrams.

The second connection is through the PX register. The PX register is 8-bits wide and can be loaded from either the lower 8 bits of the DMD bus or the lower 8 bits of the PMD bus. Its contents can also be read to the lower 8 bits of either bus.

External Buses

The program memory address bus (PMA) and the data memory address (DMA) are multiplexed into one bus and driven off chip. Likewise, the program memory data bus (PMD) and the data memory data bus (DMD) are multiplexed into one bus and driven off chip. The sixteen MSBs of the external data bus are used as the DMD bus.

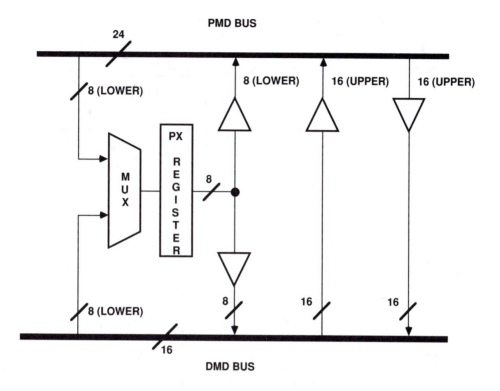

Figure 1-11: PMD-DMD Bus Exchange

1.3 ADSP-2101 BASIC SYSTEM

Figure 1-12 shows a basic system configuration with the ADSP-2101, two serial devices, a boot EPROM and optional external program and data memories. Up to 15K words of data memory and 16K words of program memory can be supported. Programmable wait state generation allows the processor to interface easily to slow memories. In this section, we discuss interfacing of the ADSP-2101 with the external devices.

1.3.1 System Interface

In this section, we present several issues related to clock crystal and interrupt handling.

Clock Signals

The ADSP-2101 takes a TTL-compatible clock signal, CLKIN, running at the instruction rate. A clock output (CLKOUT) signal is generated by the processor synchronized to the processor's internal cycles. The rising edge of CLKOUT is aligned with the rising edge of CLKIN. CLKIN may not be halted, changed during operation or operated below the specified frequency.

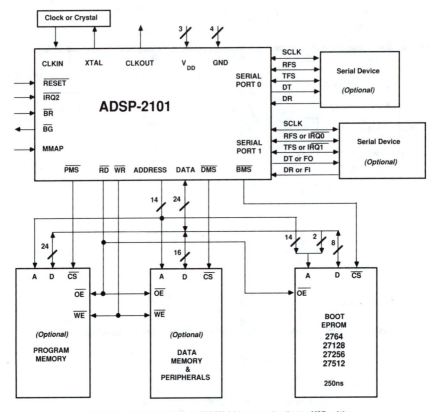

NOTE: The two MSBs of the Boot EPROM Address are also the two MSBs of the Data Bus. This is only required for the 27256 and 27512.

Figure 1-12: ADSP-2101 Basic System Configuration

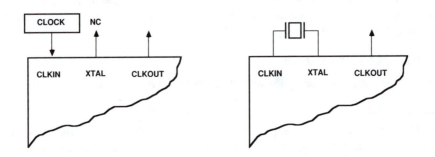

Figure 1-13: External Crystal Connections

Because the ADSP-2101 contains an internal oscillator, an external crystal may be used in place of an external clock oscillator. The crystal should be connected across the CLKIN input and the XTAL input, as shown in Figure 1-13. If an external clock oscillator is used, the XTAL input must not be connected.

Interrupt Handling

The ADSP-2101 provides up to three external interrupt input pins, $\overline{IRQ0}$, $\overline{IRQ1}$ and $\overline{IRQ2}$. $\overline{IRQ2}$ is always available as a dedicated pin; $\overline{IRQ1}$ and $\overline{IRQ0}$ may be alternately configured as part of serial port 1. The input pins can be programmed to be either level- or edge-sensitive. The ADSP-2101 also supports internal interrupts from the timer and the two serial ports. The interrupt levels are internally prioritized and individually maskable. The priorities of all six interrupts are shown below.

The ADSP-2101 supports a vectored interrupt scheme: when an interrupt is acknowledged, the processor shifts program control to the interrupt vector address corresponding to the interrupt level. Interrupts can optionally be nested so that a higher priority interrupt can preempt the currently executing interrupt service routine. Each interrupt vector location is four instructions in length, so that simple service routines can be coded entirely in this space. Longer routines require an additional JUMP or CALL.

Source of Interrupt	Interrupt Vector
$\overline{IRQ2}$ (external pin)	0004 (*highest priority*)
SPORT0 Transmit (internal)	0008
SPORT0 Receive (internal)	000C
SPORT1 Transmit (internal) or $\overline{IRQ1}$ (external)	0010
SPORT1 Receive (internal) or $\overline{IRQ0}$ (external)	0014
Timer (internal)	0018 (*lowest priority*)

$\overline{RESET}$ Signal

The RESET signal initiates a master reset of the ADSP-2101. The RESET signal must be asserted when the chip is powered up to assure proper initialization. RESET during initial power-up must be held long enough to allow the internal clock to stabilize. If RESET is activated subsequently, the clock continues and does not require this stabilization time.

The master reset sets all internal stack pointers to the empty stack condition, masks all interrupts and clears the MSTAT register. When RESET is released, if there is no pending bus request and the chip is configured for booting, the boot-loading sequence is performed. Then the first instruction is fetched from program memory location 0x0000.

Flag In and Flag Out Pins

In addition to the $\overline{IRQ1}$ and $\overline{IRQ2}$ pins, the alternate configuration of SPORT1 provides the ADSP-2101 with a Flag In (FI) and a Flag Out (FO) pin. In the alternate configuration, the DR1 pin is redefined as Flag In and the DT1 pin as Flag Out. Clearing the SPORT1 configuration bit in the system control register selects the alternate configuration.

FI may be used to control the branching of program. FO may be set, toggled, or cleared in software to signal events or conditions to any other device such as a host processor. FO is available as a read-only bit of the SPORT1 control register.

1.3.2 Memory Interface

In this section, we describe bus interconnections and program/data memory maps as well as interfaces.

Program Memory Interface

The program memory address bus (PMA) and the program memory data bus (PMD) are multiplexed with DMA and DMD, sharing the external data and address bus. The 14-bit address bus directly addresses up to 16K words, of which 2K are on-chip. The data bus is bidirectional and 24 bits wide to external program memory.

There is no placement restriction for instruction code and data in the program memory space, except for the locations used for interrupt and restart vectors.

The program memory data lines are bidirectional. The Program Memory Select ($\overline{\text{PMS}}$) signal indicates access to the Program Memory and can be used as a chip select signal. The Write ($\overline{\text{WR}}$) signal indicates a write operation and can be used as a write strobe. The Read ($\overline{\text{RD}}$) signal indicates a read operation and can be used as a read strobe or output enable signal.

The ADSP-2101 writes data from its 16-bit registers to the 24-bit program memory using the PX register to provide the lower eight bits. When it reads data (not instructions) from 24-bit program memory to a 16-bit data register, the lower eight bits are placed in the PX register.

Program Memory Maps

Program memory can be mapped in two ways, depending on the state of the MMAP pin. Figure 1-14 shows the two configurations. When MMAP=0, the internal RAM occupies 2K words beginning at address 0x0000. The external program memory uses the remaining 14K words beginning at address 0x0800. In this configuration, the boot loading sequence (described below) is automatically initiated when $\overline{\text{RESET}}$ is released.

When MMAP=1, 14K words of external program memory begin at address 0x0000 and internal RAM is located in the upper 2K words, beginning at address 0x3800. In this configuration, program memory is not loaded although it can be written to and read from under program control.

Boot Memory Interface

The Boot memory space consists of an external 64K by 8 space, divided into eight separate 8K by 8 pages. Three bits in the system control register select which page is loaded by the Boot memory interface. Another bit in the system control register allows the user to force a boot loading sequence under software control. Boot loading from page 0 after $\overline{\text{RESET}}$ is initiated automatically if MMAP=0.

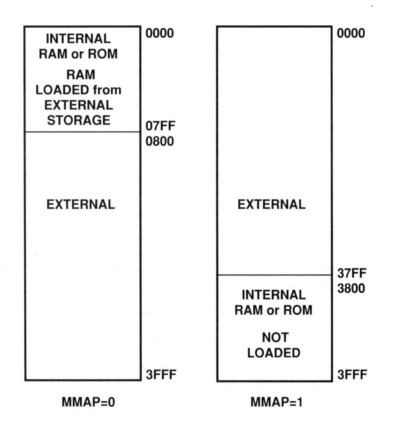

Figure 1-14: ADSP-2101 Program Memory Map

The boot memory interface can generate 0 to 7 wait states; it defaults to 3 wait states after RESET. This allows a 50MHz processor to use a slow (250 ns), low-cost EPROM for program storage. Program memory is loaded a byte at a time and converted to 24-bit words.

The $\overline{\text{BMS}}$ and $\overline{\text{RD}}$ signals are used to select and strobe the boot memory interface. Only 8-bit data is read over the data bus. To accommodate up to eight pages of boot memory, the two MSBs of the data bus are used in the boot memory interface as the two MSBs of the boot space address.

$\overline{\text{BR}}$ is recognized during the booting sequence. The bus is granted after the completion of loading the current byte. $\overline{\text{BR}}$ during booting may be used to implement booting under the control of a host processor.

Data Memory Interface

The data memory address bus (DMA) and the data memory data bus (DMD) are multiplexed with PMA and PMD, sharing the external data and address bus. The DMA bus is 14 bits wide. The bidirectional external data bus is 24 bits wide, with the upper 16 bits used for DMD transfers.

The Data Memory Select ($\overline{\text{DMS}}$) signal indicates access to the Data Memory and can be used as a chip select signal. The Write ($\overline{\text{WR}}$) signal indicates a write operation and can be used as a write strobe. The Read ($\overline{\text{RD}}$) signal indicates a read operation and can be used as a read strobe or output enable signal.

The ADSP-2101 supports memory-mapped I/O, with the peripherals memory mapped into the data memory address space and accessed by the processor in the same manner as data memory.

Data Memory Map

The on-chip data memory RAM resides in the 1K words of data memory beginning at address 0x3800, as shown in Figure 1-15. In addition, data memory locations from H#3C00 to the end of data memory at 0x3FFF are reserved. Control registers for the system, timer, wait state configuration and serial port operations are located in this region of memory.

The remaining 14K of data memory is external. External data memory is divided into five zones associated with five different wait states. This allows slower peripherals to be mapped into zones of data memory with more wait states. Figure 1-15 shows these zones.

Bus Interface

The ADSP-2101 can relinquish control of the data and address buses to an external device. When the external device requires access to memory, it asserts the Bus Request ($\overline{\text{BR}}$) signal. If the ADSP-2101 is not performing an external memory access, then it responds to the active BR input in the same cycle by:

- tristating the Data and Address bus and the $\overline{\text{PMS}}$, $\overline{\text{DMS}}$, $\overline{\text{BMS}}$, $\overline{\text{RD}}$, $\overline{\text{WR}}$ output drivers,
- asserting the Bus Grant ($\overline{\text{BG}}$) signal,

- completing the current instruction, and
- halting program execution.

If the Go mode is set, however, the ADSP-2101 will not halt program execution until it encounters an instruction that requires an external memory access.

If the ADSP-2101 is performing an external memory access when the external device asserts the $\overline{\text{BR}}$ signal, then it will not tristate the memory interfaces or assert the $\overline{\text{BG}}$ signal until the cycle after the access completes, up to eight cycles later depending on the number of wait states. The instruction does not need to be completed when the bus is granted; the ADSP-2101 will grant the bus in between two memory accesses if an instruction requires more than one memory access.

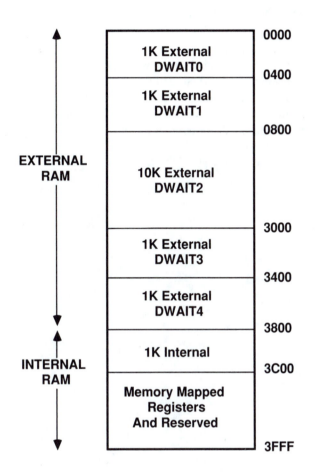

Figure 1-15: ADSP-2101 Data Memory Map

When the $\overline{\text{BR}}$ signal is released, the processor releases the $\overline{\text{BG}}$ signal, re-enables the output drivers and continues program execution from the point where it stopped.

1.4 SUMMARY

In this chapter we briefly described the architecture of the ADSP-2101 microcomputer. It exhibits a high degree of parallelism, tailored to DSP requirements. The key features of this architecture are:

- 12.5 MHz instruction rate, 80ns per instruction.
- Single-cycle access to both program and data memory.
- High degree of parallelism to:

- — compute the next program address
- — fetch the next instruction
- — perform one two data moves
- — update two address pointers
- — perform a computation
- — receive and transmit data via two serial ports.
- Code compatibility with ADSP-2100.

In the next chapter we present an overview of the ADSP-2101 instruction set. The basic understanding of architecture is necessary and useful in learning intricate details of assembly language programming using the instruction set.

chapter 2

ADSP-2101 INSTRUCTION SET OVERVIEW

2.1 INTRODUCTION

This chapter provides an overview of the instruction set used to program the ADSP-2101 chip. It provides enough information to understand the nature of programming the ADSP-2101 and the capabilities of the instruction set itself. This chapter is *not* a programmer's reference and therefore the ADSP-2101 *Cross-Reference Manual* [2] must be consulted for a complete reference to the instruction set.

For software development, the programmer must have access to the development tools: System Builder, Assembler, Linker, Simulator and PROM Splitter. The overview of the ADSP-2101 development system software is given in the next chapter. The complete description of these tools is also given in the Cross-Software Manual.

The ADSP-2101 instruction set is tailored to the computation-intensive algorithms common in DSP applications. This is possible because the instruction set allows data movement between various computational units with minimum overhead. For example, sustained single-cycle multiplication/accumulation operations are possible. The instruction set provides full control of the ADSP-2101's three computational units: the ALU, MAC and Shifter. Arithmetic instructions can process single precision 16-bit operands directly with provisions for multiprecision operations. The ADSP-2101 assembly language uses an algebraic syntax for arithmetic operations and for data moves resulting in highly readable source code. The sources and destinations of computations and data moves are written explicitly, eliminating cryptic assembler mnemonics. There is no performance penalty for this; each program statement assembles into one 24-bit instruction which executes in one cycle. There are no multicycle instruction in the ADSP-2101 instruction set. Some fifty registers surrounding the computational units are dual purpose: they are available for general purpose on-chip storage when not used in computation. This saves many memory access cycles and provides excellent freedom in coding.

The control instructions provide conditional execution of most calculations and, in addition to usual the JUMP and CALL, supports a DO UNTIL looping instruction. Return from interrupt (RTI) and the return from subroutine (RTS) are also provided. These services are made compact and speedy by the single cycle context save. The contents of the primary register set are held constant while the alternate set is enabled for subroutine and interrupt services. This eliminates the cluster of PUSHes and POPs of stacks common in general purpose microprocessors.

The ADSP-2101 also provides the IDLE instruction for idling the processor until an interrupt occurs. IDLE puts the processor into a low-power state while waiting for interrupts. Two addressing modes are supported for memory fetches. Direct addressing uses immediate values; indirect addressing uses the two Data Address Generators (DAGs).

The 24-bit instruction word allows a high degree of parallelism in performing operations. The instruction set allows for a single-cycle execution of any of the following combinations:
- any ALU, MAC or Shifter operation (may be conditional)
- any register to register move
- any data memory read or write
- a computation with any data register to data register move
- a computation with any memory read or write
- a computation with a read from two memories.

Symbol	Meaning
+,-	Add, Subtract
*	Multiply
a=b	Transfer into *a* the contents of *b*
,	Separates multifunction instructions
DM(addr)	The contents of data-memory at location "addr"
PM(addr)	The contents of program-memory at location "addr"
[Option]	Anything within aquare brackets is an optional part of the instruction statement
| option a | | option b |	List of parameters enclosed by parallel vertical lines require the choice of one parameter from among the available list.
CAPITAL LETTERS	Denote reserved words. These are instruction words, register names and operand selections.
parameters	are shown in small letters and denote an operand in the instruction for which there are numerous choices.
<data>	denotes an immediate data value.
<addr>	denotes an immediate value of an address to be coded in the instruction.
;	End of instruction.

Table 2-1: Notation Used in Instruction Set ———————————————

The ADSP-2101 instruction set provides the programmer with maximum flexibility. The instruction set provides moves from any register to any other register, or from most registers to/from either memory. For combining operations, almost any ALU, MAC or Shifter operation may be combined with any register-to-register move or with a register move to or from either internal or external memory.

There are five basic categories of instructions: computational instructions, data move instructions, multifunction instructions, program flow control instructions and miscellaneous instructions. Each of these instruction types is described in the next several sections. At the end of each section, tables summarizing the syntax of each instruction category is given. The notation used in an instruction is shown in Table 2-1.

2.2 COMPUTATION INSTRUCTIONS

This group of commands execute all ALU, MAC and Shifter instructions. There are two functional classes: *standard* instructions, which include the bulk of the computation operations, can be executed conditionally (IF condition…) which test the ALU status register, and may be combined with a data transfer in single-cycle multifunction instructions; and *special* instructions which form a small subset and must be executed individually. The permissible conditions are listed in Table 2-2.

Condition	Keyword
ALU result is:	
equal to zero	EQ
not equal to zero	NE
greater than zero	GT
greater than or equal to zero	GE
less than zero	LT
less than or equal to zero	LE
ALU carry status:	
carry	AC
not carry	NOT AC
x-input sign:	
positive	POS
negative	NEG
ALU overflow status:	
overflow	AV
not overflow	NOT AV
MAC overflow status:	
overflow	MV
not overflow	NOT MV
Counter status:	
not expired	NOT CE

Table 2-2: Permissible Conditions for Computation Instructions

    Each computational unit has a set of input registers and output registers. A list of permissible input operands and result registers are given in Table 2-3.

ALU

Source for X input (xop)	Source for Y input (yop)	Destination for output port R
AX0, AX1, AR MR0, MR1, MR2 SR0, SR1	AY0, AY1 AF	AR AF

MAC

Source for X input (xop)	Source for Y input (yop)	Destination for output port R
MX0, MX1, AR MR0, MR1, MR2 SR0, SR1	MY0, MY1 MF	MR (MR2, MR1, MR0) MF

Shifter

Source for Shifter input (xop)	Destination for Shifter output
SI, SR0, SR1 AR MR0, MR1, MR2	SR (SR1, SR0)

Table 2-3: Computational Input/Output Registers ———————————

2.2.1 ALU Group

Standard Functions: Standard ALU instructions include add, subtract, logic (AND, OR, NOT, eXclusive-OR), pass, negate, increment, decrement, clear, and absolute value. The "-" function does twos-complement subtraction while NOT obtains a ones-complement. The PASS function passes the listed operand but tests and stores status information for later sign/zero testing. As an example, consider an ALU addition instruction for add/add-with-carry in the form:

```
[IF condition] | AR |    =    xop  | + yop     | ;
               | AF |              | + C       |
                                   | + yop + C |
```

Instructions are in similar form for subtraction and logical operations. If the options AR and "+ yop + C" are chosen, and if xop and yop are the contents of AX0 and AY0 respectively, the unconditional instruction would read:

```
AR = AX0 + AY0 + C;
```

This algebraic expression means that the ALU result register AR gets the value of the ALU x-input and y-input registers plus the value of the carry-in bit. This shortens code and speeds execution by eliminating many separate register-move instructions.

When an optional IF condition is included, and if ALU Carry bit status is chosen then the conditional instruction would read:

```
IF AC AR = AX0 + AY0 + C;
```

The conditional expression, IF AC, tests the ALU Carry bit; if there is a carry from the previous instruction, this instruction executes, otherwise a NOP occurs and execution continues with the next instruction.

Special Functions: The division instruction is the only ALU special function. It is executed in two steps: DIVS computes the sign, then DIVQ computes the quotient. A full divide of a signed 16-bit divisor into a signed 32-bit quotient requires a DIVS followed by 15 DIVQ's.

Table 2-4 is a list of all ALU instructions.

[IF condition]	AR / AF	=	xop	+ yop + C + yop + C	;
[IF condition]	AR / AF	=	xop	− yop − yop + C − 1	;
[IF condition]	AR / AF	=	yop	− xop − xop + C − 1	;
[IF condition]	AR / AF	=	xop	AND / OR / XOR yop	;
[IF condition]	AR / AF	=	PASS	xop / yop / 0 / 1	;
[IF condition]	AR / AF	=	−	xop / yop	;
[IF condition]	AR / AF	=	NOT	xop / yop	;
[IF condition]	AR / AF	=	ABS	xop	;
[IF condition]	AR / AF	=	yop	+ 1	;
[IF condition]	AR / AF	=	yop	− 1	;
DIVS yop, xop ;					
DIVQ xop ;					

Table 2-4: ALU Instructions

2.2.2 MAC Group

Standard Functions: Standard MAC instructions include multiply, multiply accumulate, multiply-subtract, transfer AR conditionally, and clear. As an example, consider a MAC instruction for multiply-accumulate in the form:

```
[IF condition] | MR |     =       MR + xop * yop ( | SS | );
               | MF |                              | SU |
                                                   | US |
                                                   | UU |
                                                   | RND|
```

If the options "MR" and "UU" are chosen, if xop and yop are the contents of MX0 and MY0 respectively, and if MAC overflow condition is chosen, then a conditional instruction would read:

```
        IF NOT MV MR = MR + MX0*MY0 (UU);
```

The conditional expression, IF NOT MV, tests the MAC overflow bit. If the condition is not true, a NOP is executed. The expression MR=MR+MX0*MY0 is the multiply/accumulate operation: the multiplier result register (MR) gets the value of itself plus the product of the X and Y input registers selected. The modifier in parentheses (UU) treats the operands as unsigned. There can be only one such modifier selected from the available set. (SS) means both are signed, while (US) and (SU) mean that either the first or second operand is signed; (RND) means to round the (implicitly signed) result.

Special Functions: Accumulator saturation is the only MAC special function.

```
        IF MV SAT MR;
```

The instruction tests the MAC overflow bit (MV) and saturates the MR register (for only one cycle) if that bit is set.

Table 2-5 is a list of all MAC instructions.

```
[IF condition] | MR |     =     xop * yop ( | SS | );
               | MF |                       | SU |
                                            | US |
                                            | UU |
                                            | RND|

[IF condition] | MR |     =     MR + xop * yop ( | SS | );
               | MF |                            | SU |
                                                 | US |
                                                 | UU |
                                                 | RND|

[IF condition] | MR |     =     MR - xop * yop ( | SS | );
               | MF |                            | SU |
                                                 | US |
                                                 | UU |
                                                 | RND|

[IF condition] | MR |     =     0;
               | MF |
```

```
[IF condition]  | MR  |      =    MR [( RND )];
                | MF  |

IF MV SAT MR;
```

Table 2-5: MAC Instructions ————————————————————————————

2.2.3 Shifter Group

Standard Functions: Shifter standard functions include arithmetic and logical shifts, as well as floating point and block floating point scaling operations; derive exponent, normalize, denormalize, and block exponent adjust. As an example, consider a Shifter instruction for normalize:

```
IF NOT CE SR = SR OR NORM SI (HI);
```

The conditional expression, IF NOT CE, tests the "not counter expired" condition. If the condition is false, a NOP is executed. The destination of all shifting operations is the Shifter Result register, SR. (The destination of the exponent detection instructions is SE or SB, as shown below.) In this example, SI, the Shifter Input register, is the operand. The amount and direction of the shift is controlled by the signed value in the SE register in all shift operations except an immediate shift. Positive values cause left shifts; negative values cause right shifts.

The "SR OR" modifier (which is optional) logically ORs the result with the current contents of the SR register; this allows the user to construct a 32-bit value in SR from two 16-bit pieces. "NORM" is the operator and "(HI)" is the modifier that determines whether the shift is relative to the HI or LO (16-bit) half of SR. If "SR OR" is omitted, the result is passed directly into SR.

Special Functions: Shift-immediate is the only Shifter special function. The number of places (exponent) to shift is specified in the instruction word.

Table 2-6 provides a list of all Shifter instructions.

```
[IF condition]  SR    =    [SR OR] ASHIFT  xop ( | HI |  );
                                                 | LO |

[IF condition]  SR    =    [SR OR] LSHIFT  xop ( | HI |  );
                                                 | LO |

[IF condition]  SR    =    [SR OR] NORM    xop ( | HI |  );
                                                 | LO |

[IF condition]  SE    =    EXP             xop ( | HI  |  );
                                                 | LO  |
                                                 | HIX |

[IF condition]  SE    =    EXPDJ           xop ;

SR    =    [SR OR] ASHIFT  xop BY <data> ( | HI |  );
                                           | LO |

SR    =    [SR OR] LSHIFT  xop BY <data> ( | HI |  );
                                           | LO |
```

Table 2-6: Shifter Instructions ————————————————————————

2.3 DATA MOVE INSTRUCTIONS

These instructions move data to and from data registers and external memory. ADSP-2101 registers are divided into two groups, referred to as *reg* which includes almost all registers and *dreg* or data registers, which is a subset. Only the program counter (PC) and the ALU and MAC feedback registers (AF and MF) are not accessible. Table 2-7 shows which registers belong to these groups. Many of the ADSP-2101 system control registers are memory-mapped. These are read and written as memory locations instead of with register names.

Accessible Registers: reg

	Data Registers: dreg
SB	AX0, AX1, AY0, AY1, AR
PX	MX0, MX1, MY0, MY1, MR0, MR1, MR2
I0 - I7, M0 - M7, L0 - L7	SI, SE, SR0, SR1
CNTR	
ASTAT, MSTAT, SSTAT	
IMASK, ICNTL	
TX0, TX1, RX0, RX1	
IFC	

Table 2-7: ADSP-2101 Register Set: reg & dreg ──────────────

There are five classes of data move instruction. Except for immediate instructions, data addresses are computed by the DAGs via the contents of their index (I) and modify (M) registers. The data move classes are:

- Load register immediate
- Register-to-register move
- Immediate address DM move
- Indirect address DM or PM move
- Multifunction DM and PM read

In the description of each these classes below, "immediate value" refers to a 16-bit number contained in the instruction field while "immediate address" refers to a 14-bit address contained in the instruction field.

Load Register Immediate: In this instruction, the data is provided by a 16-bit immediate value in the instruction-word and is moved into *reg*.

        reg = <data>;   <data> = immediate value.

Register-to-Register Move: Here the value of a permissible *reg* is moved into another permissible *reg*.

        reg = reg;

Immediate Address Data Memory Move: Here data memory addressed by a 14-bit immediate address in the instruction-word is moved into a *reg*. Since no registers are tied up in generating addresses, any accessible *reg* can be the operand.

Data Memory Read
```
reg = DM(<addr>);                            <addr> = immediate address.
```

Data Memory Write
```
DM(<addr>) = reg;
```

Indirect Address Memory Move: These instructions cause data to be moved between *dreg* and either DM or PM. The memory address for the current operations is provided by one of the four I-registers (Im) of a DAG. The value of I is then stored back again after being modified by the contents of one of four M (Mn) and the buffer length stored in one of four L registers (Lm). The operation is:

Read/write at address specified by Pointer:
```
Memory address = Im
```
Then Modify pointer:
```
Im = (Im + Mn) mod Lm
```

Indirect addressing may access either data-memory or program-memory. The register set in DAG1 can only be used for data-memory while register set in DAG2 can be used for either data- or program-memory. The buffer length registers Lm are paired with the index registers Im, i.e., L5 is the modulus register for the memory location indexed at I5.

Memory Read into Data Register
```
dreg = DM(Im,Mn);
dreg = PM(Im,Mn);
```

DM Write: Immediate or from Data Register
```
DM(Im,Mn) = | dreg |
              |<data>|
```

PM Write from Data Register
```
PM(Im,Mn) = dreg;
```

Multifunction Data- and Program- Memory Read: Here a combined Move instruction reads data into a pair of ALU or MAC input registers from both DM and PM in the same cycle. It is described in more detail in the next section.

Table 2-8 gives a list of all data Move instructions.

```
reg      =      reg;
reg      =      DM (<address>);
```

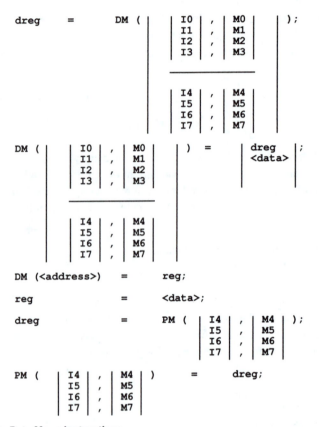

Table 2-8: Data Move Instructions ――――――――――――――――――――――――――――――

2.4 MULTIFUNCTION INSTRUCTIONS

Multifunction operations exploit the inherent parallelism of the ADSP-2101 architecture by providing combinations of data moves, memory reads and memory writes and computation in a single-cycle.

2.4.1 ALU/MAC with Data and Program Memory Read

Perhaps the most common single operation in DSP algorithms is the sum of products, like the following:

- Fetch two operands (such as a coefficient and a data point)
- Multiply them and sum the result with previous products

The ADSP-2101 can execute both data fetches and the multiplication/accumulation in a single-cycle. Typically, a loop of multiply/accumulates can be expressed in ADSP-2101 source code in just two program lines. Since the on-chip program memory is fast enough to provide an operand and the next instruction in a single cycle, loops of this type can execute with sustained single-cycle throughput. An example of such an instruction is:

MR=MR+MX0*MY0(SS), MX0=DM(I0,M0), MY0=PM(I4,M5);

The first clause of this instruction (up to the first comma) says that MR, the MAC result register, gets the sum of its previous value plus the product of the (current) X and Y input registers of the MAC (MX0 and MY0) both treated as signed (SS). Note the simple assignment statement form of the source code.

In the second and third clauses of this multifunction instruction, two new operands are fetched. One is fetched from the data memory (DM) pointed to by index register zero (I0, post modified by the value in M0) and the other is fetched from the program memory location (PM) pointed to by I4 (post-modified by M5 in this instance). Note that indirect memory addressing uses a syntax similar to array indexing, with DAG registers providing the index values. Any I register may be paired with any M register within the same DAG.

As discussed in Chapter 1, registers are read at the beginning of the cycle and written at the end of the cycle. The operands present in the MX0 and MY0 registers at the beginning of the instruction cycle are multiplied and added to the MAC result register, MR. The new operands fetched at the end of this same instruction overwrite the old operands after the multiplication has taken place and are available for computation on the following cycle. The user may, of course, load any data registers in conjunction with the computation, not just the MAC registers with a MAC operation as in our example.

The computational part of this multifunction instruction may be any unconditional ALU instruction except division or any MAC instruction except saturation. Certain other restrictions apply: the next X operand must be loaded into MX0 from data memory and the new Y operand must be loaded into MY0 from program memory (internal and external memory are identical at the level of the instruction set). The result of the computation must go to the result register (MR or AR) not to the feedback register (MF or AF).

2.4.2 Data and Program Memory Read

This instruction is a special case of the instruction above, in which the computation is left out. It is also discussed in Section 2.3 as a multifunction data move instruction. It executes only the dual fetch as shown below.

```
AX0 = DM(I2,M0), AY0=PM(IM,M6);
```

In this example, we have used the ALU input registers as the destination. As with the previous multifunction instruction, X operands must come from data memory and Y operands from program memory (internal or external memory in either case).

2.4.3 Computation With Memory Read

If a single memory read is performed, instead of the dual memory read of the previous two multifunction instructions, a wider range of computations can be executed. The legal computations include all ALU operations except division, all MAC operations and all Shifter operations except SHIFT IMMEDIATE. Computation must be unconditional.

An example of this instruction is:

```
AR=AX0+AY0, AX0=DM(I0,M3);
```

Here an addition is performed in the ALU while a single operand is fetched from data memory. The restrictions are similar to those for previous multifunction instructions. The value of AX0, used as a source for the computation, is the value at the beginning of the cycle. The data read operation loads a new value into AX0 by the end of the cycle. For this same reason, the destination register (AR in the example above) cannot be the destination for the memory read. If that were legal, there would be a conflict.

2.4.4 Computation With Memory Write

The computation with the memory write instruction is similar in structure to the immediately preceding one: the order of the clauses in the instruction line, however, is reversed. First the memory write is performed; then the computation is performed as shown below:

```
DM(I0,M0)=AR, AR=AX0+AY0;
```

Again, the value of the source register for the memory write (AR in the example) is the value at the beginning of the instruction. The computation loads a new value into the same register; this is the value in AR at the end of this instruction. Reversing the order of the clauses of the instruction is illegal and invokes an assembler warning; it would imply that the result of the computation is written to memory when, in fact, the previous value of the register is what is written. There is no requirement that the same register be used in this way although this will usually be the case in order to pipeline operands to the computation.

The restrictions on computation operations are identical to those above. All ALU operations except division, all MAC operations and all Shifter operations except SHIFT IMMEDIATE are legal. Computation must be unconditional.

2.4.5 Computation With Data Register Move

This final multifunction instruction performs a data register to data register move in parallel with a computation. Most of the restrictions applying to the previous two instructions apply to this instruction.

```
AR=AX0+AY0, AX0=MR2;
```

Here an ALU addition operation occurs while a new value is loaded into AX0 from MR2. As before, the value of AX0 at the beginning of the instruction is the value used in the computation. The move may be from or to all ALU, MAC and Shifter input and output registers except the feedback registers (AF and MF) and SB.

In the example, the data register move loads the AX0 register with the new value at the end of the cycle. All ALU operations except division, all MAC operations and all Shifter operations except SHIFT IMMEDIATE are legal. Computation must be unconditional.

A complete list of multifunction instructions appears in Table 2-9.

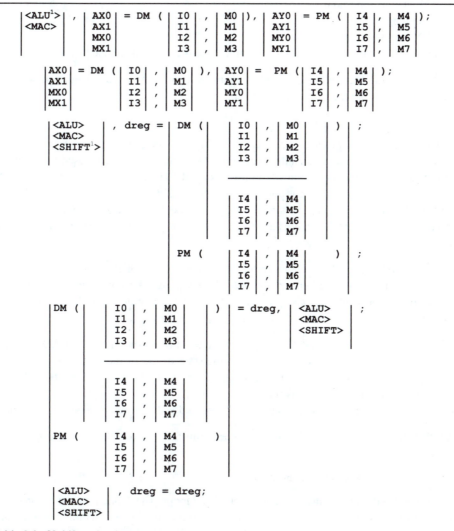

Table 2-9: Multifunction Instructions————————————————————————

2.5 PROGRAM FLOW CONTROL INSTRUCTIONS

Program flow control on the ADSP-2101 is simple yet powerful. It directs the program sequencer. In a normal order, the sequencer automatically fetches the next contiguous instruction for execution. This flow can be altered by these instructions. Program flow control provides:

- Jumps to interrupt service routines and calls to subroutines
- Return from interrupts and subroutines

---

1 All computation is unconditional; ALU Division and Shift Immediate operations prohibited.

- FLAG pin conditions
- DO loops
- IDLE instruction

An optional "IF" condition can first test any of the status conditions defined earlier.

JUMP and CALL Instructions: JUMP is a familiar construct from many other processors. As an example consider the following statement:

```
IF EQ JUMP my_label;
```

My_Label is any identifier used as a new address where the program control is transferred for execution. Instead of the label, an index register in DAG2 may be explicitly used. The default scope for any label is the module in which it is declared. The Assembler directive .ENTRY makes a label "visible" as an entry point for routines outside the module. Conversely, the .EXTERNAL directive makes it possible to use a label declared in another module. On the other hand, a CALL instruction brings in subroutines, and pushes the present PC onto its stack as the return address.

RETURN instructions: There are two return statements: RTS from subroutine, and RTI from an interrupt service routine. In either case, a RETURN pops the return address from the PC stack. A return from an interrupt also pops the status stack back, returning arithmetic and mode status and interrupt mask registers to the values they had prior to the interrupt.

FLAG Instructions: JUMP and CALL permit the additional conditionals "FLAG_IN" and "NOT FLAG_IN" to be used for branching on the state of the FI pin, but only with direct addressing, not with DAG2 as the address source. Additionally, FO pin, (Flag Out) can be set, cleared or toggled. Although this instruction does not alter the flow of the program, it provides a control structure for multiprocessor communication and is therefore included in this group.

Loop Instruction The DO UNTIL instruction performs the zero-overhead looping operation. It has the form:

```
DO <addr> [UNTIL condition];
```

The label <addr> designates the last instruction in the loop. The "condition" determines the termination of the loop. When the DO is entered, <addr> and the termination condition are pushed onto the Sequencer loop stack, and the current PC+1 is pushed onto the PC stack to become the next address after loop termination. The looping hardware automatically checks the condition code whenever execution passes through the address-value <addr>.

IDLE Instruction: This instruction provides a way to wait for interrupts. IDLE causes the processor to wait in a low-power state until an interrupt occurs. It uses less power than loops created with JUMP.

Table 2-10 provides a complete list of program flow control instructions.

[IF condition]	JUMP	(I4)	;
		(I5)	
		(I6)	
		(I7)	
		<address>	

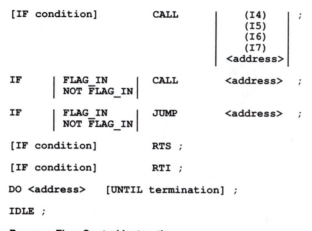

```
[IF condition]         CALL        |   (I4)   |  ;
                                   |   (I5)   |
                                   |   (I6)   |
                                   |   (I7)   |
                                   |<address>|

IF        | FLAG_IN     |  CALL     <address>   ;
          | NOT FLAG_IN |

IF        | FLAG_IN     |  JUMP     <address>   ;
          | NOT FLAG_IN |

[IF condition]         RTS ;

[IF condition]         RTI ;

DO <address>    [UNTIL termination] ;

IDLE ;
```

Table 2-10: Program Flow Control Instructions ─────────────────────────

2.6 MISCELLANEOUS INSTRUCTIONS

There are several miscellaneous instructions. NOP, of course, is a no operation instruction. The PUSH/POP instruction allows explicit control of the status, counter, PC and loop stacks; interrupt servicing automatically pushes and pops some of these stacks.

The Mode Control (enable/disable) instructions turn on and off several modes of operation. The instruction governs modes common to the ADSP-2100 (bit-reversal on DAG1, latching ALU overflow, saturating the ALU result register, choosing the primary or shadow register set) and the ADSP-2101 extended mode controls (GO mode for continued operation during Bus Grant, multiplier shift mode for fractional or integer arithmetic and timer enabling).

A single ENA or DIS can be followed by any number of mode identifiers, separated by commas; ENA and DIS can also be repeated. All seven modes can be enabled, disabled or changed in a single instruction.

The MODIFY instruction modifies the address pointer in the I register selected with the value in the selected M register, without performing any actual memory access. As always, the I and M registers must be from the same DAG; any of I0-I3 may be used only with one from M0-M3 and the same for I4-I7 and M4-M7. If circular buffering is in use, modulus logic applies (See Chapter 3, "Data Moves," for more information).

Table 2-11 gives a complete list of miscellaneous instructions.

```
[IF condition]     | SET    |     FLAG_OUT ;
                   | RESET  |
                   | TOGGLE |

NOP;
```

```
| PUSH |          STS [, POP CNTR] [, POP PC] [, POP LOOP] ;
| POP  |

| ENA |           | BIT_REV  |   [ , ... ] ;
| DIS |           | AV_LATCH |
                  | AR_SAT   |
                  | SEC_REG  |
                  | TIMER    |
                  | G_MODE   |

MODIFY    (       | I0 |   ,   | M0 |           ) ;
                  | I1 |       | M1 |
                  | I2 |   ,   | M2 |
                  | I3 |   ,   | M3 |
                  _____


                  | I4 |   ,   | M4 |
                  | I5 |   ,   | M5 |
                  | I6 |   ,   | M6 |
                  | I7 |   ,   | M7 |
```

Table 2-11: Miscellaneous Instructions ───────────────────────────────

2.7 DATA STRUCTURES

The ADSP-2101 Cross-Software supports the declaration and use of a simple set of data structures: one-dimensional arrays and ports. The array may be a single value or multiple values. In addition, the array may be used as a circular buffer. Here is a brief discussion of each instance with an example of how they are declared and used. Complete syntax for these and other directives is given in the *ADSP-2101 Cross-Software Manual* [2].

2.7.1 Arrays

Arrays are the basic data structures in the ADSP-2101 instruction set. In ADSP-2101 literature, the words "array" and the expression "data buffer" are used interchangeably. Arrays are declared with Assembler directives and can be referenced indirectly and by name, can be initialized from immediate values in a directive or from external data files and can be linear or circular with automatic wraparound. Assembler Directives are described in detail in Chapter 3.

An array is declared with a directive such as

```
.VAR/DM coefficients[128];
```

This declares an array of 128 16-bit values located in data memory (DM). The special operators ^ and % reference the address and length, respectively, of the array. It could be referenced as shown below.

```
IO = ^coefficients          {point to address of buffer}
MX0=DM(IO,M0);              {load MX0 from buffer}
```

These instructions load a value into MX0 from the beginning of the coefficients buffer in data memory. With the automatic post-modify of the DAGs, the user could execute the second of these instructions in a loop and continuously advance through the buffer.

Alternatively, when only the first location needs to addressed, one can directly use the buffer name as a label in many circumstances, such as

```
MX0=DM(coefficients);
```

The Linker substitutes the actual address for the label. It is also possible to initialize a complete array/buffer from a data file, using the INIT directive.

```
.INIT coefficients : <filename.dat>;
```

This reads the values from the file filename.dat into the array at link time. This feature is supported only in the ADSP-210X Simulators even though data cannot be loaded directly into on-chip data memory by the hardware booting sequence.

An array or data buffer with a length of one behaves as a simple single-word variable.

2.7.2 Circular Arrays/Buffers

A common requirement in DSP is the circular buffer. This is directly implemented by the ADSP-2101 DAGs, using the L (length) registers. First, the buffer must be declared as circular:

```
.VAR/DM/CIRC coefficients[128];
```

This identifies it to the Linker for placement on the proper address boundary. Next, the L register must be initialized, typically using the % operator (or a constant) and, in the example below, the I register and M register.

```
L0 = %coefficients;          {length of circular buffer}
I0 = ^coefficients;          {point to address of buffer}
M0 = 1;                      {increment by 1 location each time}
```

Now a statement such as

```
MX0=DM(I0,M0);               {load MX0 from buffer}
```

in a loop, cycles continuously through coefficients and wraps around automatically. L registers should be initialized to zero for buffers of any length that are not circular.

2.7.3 Ports & Memory-Mapping

The .PORT directive in the System Builder module allows the user to refer to a specific hardware address with an identifier of his choosing as shown here. This capability makes it easy to interface to memory-mapped peripherals, such as converters.

```
.PORT/ABS= H#800 converter_in;
```

After declaring the same identifier in the Assembler, a value can be read directly from the port with a statement such as

```
SI = DM(converter_in);
```

This loads the SI register with the value present at the address specified in the System Builder. (The Linker reads the Architecture Description file produced by the System Builder to obtain the actual address for the label.) The user can change the hardware address of the port without having to rewrite the entire program.

2.8 SUMMARY

In this chapter, we reviewed the very rich instruction set through which the user gains access to many unusual features of the ADSP-2101 microcomputer. Some of these features are:

- Two addressing modes;
- Multifunction instructions: arithmetic and data moves in one cycle;
- Single-cycle context save: swaps the full register set;
- Zero-overhead looping;
- Parallel access to both data program operands.

The ADSP-2101 instruction set discussed in this chapter is a superset of the ADSP-2100 instruction set. It is source and object code compatible with the ADSP-2100. An ADSP-2100 program may need to be relocated to utilize internal memory and conform to the ADSP-2101's interrupt vector and reset vector placement.

A complete program example using this instruction set requires knowledge of the software development system and of Assembler directives. Therefore it is deferred until the next chapter which gives an overview of the ADSP-2101 development tools.

```
chapter 3



Overview of Development Tools
```

3.1 INTRODUCTION

The ADSP-2101 is a compact yet very efficient VLSI circuit containing 68 pins. As described in chapter 1, these pins provide terminals for power, data and program memory, input/output, as well as a number of other connections for interfacing with the outside world. There are several important issues that must be considered before the ADSP-2101 can be successfully interfaced in a real application. First, the chip itself must be placed in a proper system architecture (also known as the target hardware configuration) that supports it. This architecture contains memory resources and input/output devices. Second, the chip must be correctly programmed in its native assembly language so that it can perform its intended function. In almost all applications this source program code is not known in advance and therefore must be developed from scratch with several iterations of debugging. Third, the chip and its object code must be tested in an actual environment with real input/outputs with interrupts and real memory resources driven at the true speed of the chip. Finally, the object code must be burned in a PROM (or EPROM) and the chip along with the PROM must be integrated with the target system. This activity is called system development and is the main hidden cost of digital signal processing. It is carried out by the development tools provided by the manufacturer of the chip.

Although this system development is needed only initially when the application is being designed and not in the final product, it must be stressed that a student or an engineer will most likely be working with development tools. Therefore a complete understanding of the capabilities of these tools is as essential as the architecture of the chip itself. In this chapter, we briefly describe development tools that support the ADSP-2101 microcomputer. For detailed descriptions of these tools, appropriate manuals [2 – 4] are recommended.

The development process begins with the task of defining the target system hardware environment. To define the hardware environment, the System Builder is used. The System Specification file includes the target hardware information. The System Builder reads this file and creates an Architecture Description file which passes information about the target hardware to the Linker, Simulator, and Emulator.

Code generation begins by creating assembly source code modules. An assembly module is a unit of source code such as a calling program, subroutine, data buffer declaration section or any combination. Each assembly code module is assembled separately by the Assembler. Several modules are then linked together to form an executable program.

The Linker needs the target hardware information located in the Architecture Description file to determine placement of the code and data fragments. In the assembly modules we have the option to specify each code/data fragment as completely relocatable, relocatable within a defined memory segment, or placed at an absolute address. Absolute code or data modules are placed at the specified base address, provided the specified memory area has the correct attributes. Relocatable objects are placed in memory by the Linker.

Using the Architecture Description file and the Assembler output files, the Linker determines the placement of relocatable code and data segments (including circular buffers), and places all segments in memory locations with the correct attributes (CODE or DATA, RAM or ROM). The Linker generates an executable image file, which may be loaded into the Simulator and Emulator for debugging.

The Simulator provides windows that display different aspects of the hardware environment. To replicate the target hardware environment, the Simulator configures its memory according to the System Builder output, and simulates I/O ports according to user-entered Simulator commands. This simulation provides capabilities to debug the system and analyze performance before committing to a hardware prototype.

After debugging with the Simulator, the Emulator is used in the prototype target system to debug hardware, timing, and real-time software problems. It provides overlay memory to replace target system off-chip memory, including boot memory, if desired.

The PROM Splitter translates the executable memory image file (Linker output) into a file that is compatible with a PROM burner. Once the ADSP-2101 code is burnt into PROM and an ADSP-2101 is plugged into the target board, the prototype is ready to run.

Figure 3-1 shows a flow chart of the ADSP-2101 development cycle. All the above steps in the development process except emulation are carried out by the software development system while the hardware development consists of the Emulator and the prototype target system.

In the remainder of this chapter we explain each of these tools. The software development tools are described in Section 3.2. In Section 3.3 we present hardware development tools. Finally in Section 3.4, some issues related to the host computer are discussed.

3.2 SOFTWARE DEVELOPMENT TOOLS

The software development system of the ADSP-2101 is called the Cross-Software system. It is a set of modules. The System Builder module provides a high level method for defining the architecture of systems under development. The Assembler module produces object code and the Linker module combines object codes and library calls into an executable file. The Simulator module provides an interactive instruction-level simulation with a reconfigurable user interface. A PROM Splitter generates PROM burner compatible files. In addition, the C Compiler is also available (but not discussed in this book) which generates ADSP-2101 assembly source code.

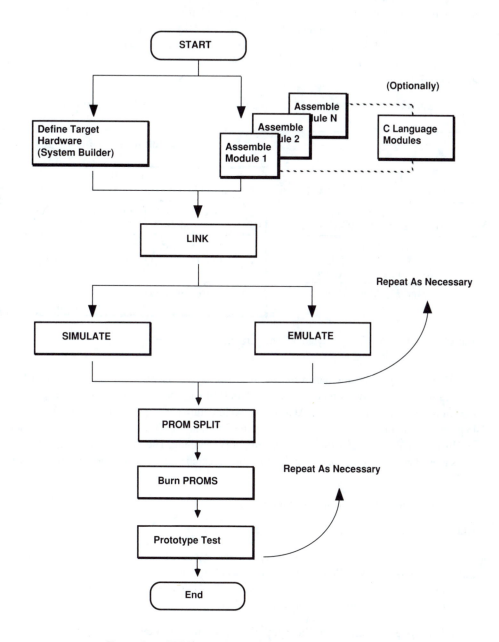

Figure 3-1: ADSP-2101 System Development Flow

3.2.1 The System Builder

The system builder module of the Cross-Software system is a software tool for describing the configuration of the target system's memory and I/O to the rest of the development system. The available absolute memory-address ranges are listed for program memory (PM) and data memory (DM) along with the intended attributes. The DM is usually RAM while the PM can be assigned to ROM for permanent programs or constants as well as to RAM blocks for coefficients or programs which may change. The memory-mapped addresses of I/O devices are also specified in this step.

The amount of memory and I/O ports a target system may include is shown in Table 3-1.

	Maximum Available
Data Memory (16-bit data, ROM or RAM)	Up to 15 K words (1K on-chip, up tp 14K off-chip, 1K reserved)
Program memory (24-bit code or data, ROM or RAM	Up to 16K words, mixed code & data (2K on-chip, up to 14K off-chip)
Boot Memory (24-bit code or data, padded to 32-bit word width)	Up to 64K bytes, configured as 16K words (1 to 8 pages, each containing 2K words)
Memory-Mapped I/O	Any number, up to memory limits (Simulator limited by host file system limits)

Table 3-1: ADSP-2101 Target System Configurations ───────────────

The user specifies the hardware configuration in a System Specification source (.SYS) file using System Builder directives. The System Builder processes the .SYS file and generates the Architecture Description file (.ACH). The Architecture Description file is used by the Linker to place relocatable segments in memory, by the Simulator to simulate memory configurations, and by the Emulator to set up target system memory mapping. The System Builder outputs error messages, if any, or a summary of the architecture created to the screen. The system builder I/O is shown in Figure 3-2.

The System Builder is invoked by typing

BLD21 *filename*[**.ext**]   [**-switch**]

where *filename.ext* is the system specification source file. The filename extension is optional and defaults to .SYS. There is one switch for invoking the System Builder. The -c switch makes the System Builder case-sensitive. This is provided primarily for compatibility with the C Compiler, which is always case-sensitive.

In a System Specification file, symbolic names are assigned to the system configuration itself, I/O ports, and memory segments. The memory segment names may be used in the Assembler; memory segment names and memory characteristics are used by the Linker.

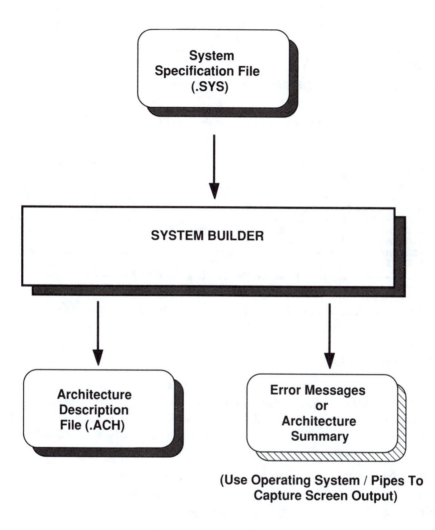

Figure 3-2: System Builder I/O

All symbolic names must be unique. A symbolic name is a string of letters, digits, and underscores with a letter as the first character. Symbol names can be of any length. Only 32 characters are significant.

System Builder keywords cannot be used as symbolic names. Table 3-2 lists the System Builder keywords.

Assembler keywords, listed in Table 3-4, may not be used as symbolic names either. The System Builder accepts such symbol definitions without flagging an error. However, the Linker does not.

ABS	CODE	ENDSYS	PORT	SYSTEM
ADSP2100	CONST	MMAP0	RAM	
ADSP2101	DATA	MMAP1	ROM	
BOOT	DM	PM	SEG	

Table 3-2: System Builder Keywords

Listing 3-1 is an example of a system specification file for an ADSP-2101 system. Note that comment fields are enclosed within braces, { }, and can be inserted anywhere in the file.

```
.SYSTEM fir_system;                                  {system name}
.ADSP2101;                                           {ADSP-2101 system}
.MMAP0;                                              {boot loading enable}
.SEG/ROM/BOOT=0 boot_mem[2048];                      {boot page one}
.SEG/PM/RAM/ABS=0/CODE/DATA int_pm[2048];            {on-chip program mem}
.SEG/PM/RAM/ABS=2048/CODE/DATA ext_pm[14336];        {external program mem}
.SEG/DM/RAM/ABS=0/DATA ext_dm[14336];                {external data mem}
.SEG/DM/RAM/ABS=14336/DATA int_dm[1024];             {on-chip data mem}
.ENDSYS;
```
———————— Listing 3-1: Sample System Specification File ——————————

The System Specification Source file for the ADSP-2101 specifies the amount of data, program, and boot memory included in the target system. The first directive in the file is the .SYSTEM directive. This directive assigns a name *fir_system* to the hardware description and signals the start of the file.

The .ADSP2101 statement identifies the processor type, here naming the ADSP-2101 microcomputer. This statement is required. The presence of the .MMAP directive or the declaration of boot memory also serves to signal the Cross-Software that the system in question is an ADSP-2101 architecture. If none of these indicators are present, the System Builder assumes an ADSP-2100 processor.

The .MMAP0 directive specifies the simulated state of the MMAP pin on the ADSP-2101 in this example system. Defining MMAP as 0 indicates that boot memory is to be loaded into the chip's internal program memory space, beginning at address 0.

The .SEG directive declares the system's physical memory segments and their characteristics. In this example, the segments declared comprise the full on-chip and off-chip program and data memory configuration of the ADSP-2101. Many applications, however, do not require this much memory space.

Boot_mem identifies a 2K-word space for one page of external boot memory.

Int_pm declares the 2K-word on-chip program memory space beginning at address 0. In the ADSP-2101 this memory can always hold both code and data and should be explicitly declared as such as in this example. *Ext_pm* declares a 14K-word space for external program code and data storage beginning at address 2048, after the on-chip memory.

Ext_dm declares a 14K-word space for external data storage beginning at address 0. *Int_dm* declares the 1K-word internal data memory space beginning at address 14336. This corresponds exactly to the on-chip data memory of the ADSP-2101 which is available for general system use. The 1K of on-chip memory above this is reserved for processor use and should not be declared. The memory segments can be declared in any order.

The last statement in a system specification file is the .ENDSYS directive. The System Builder stops processing when it encounters the .ENDSYS directive.

A complete description of each System Builder directive is given in Chapter 4. In summary, the System Builder:

- allows the user to specify target hardware,
- uses high level constructs, and
- flags inconsistencies between hardware and software.

3.2.2 The Assembler

The ADSP-2101 Assembler translates source code modules into object code modules. The user creates a source code file (.DSP) using the ADSP-2101 assembly language and define variables, data buffers, and symbolic constants using assembler directives. Separately assembled modules are linked together to form an executable program.

Figure 3-3 shows the Assembler input and output files. The ADSP-2101 Assembler reads the source code file (.DSP) and generates four output files with the same root name: an object file (.OBJ), a code file (.CDE), an initialization file (.INT), and a list file (.LST). The object file, code file and initialization files are passed to the Linker. The object file contains information on memory allocation and symbol declarations. The code file contains instruction opcodes with unresolved symbols marked. The initialization file contains initialization information for data buffers. The list file, which is optional, is for documentation.

Using assembly directives in the source code file, the user can include other source code files and inform the Linker of initialization data files in the assembly process. The Assembler reads these files and processes them together with the original source file. There are two pre-processors of the Assembler, an ANSI-standard C language module and a standard preprocessor. The Assembler also supports a macro capability.

Assembler Modules

The Assembler consists of three modules:

C language preprocessor	actual filename: ASMPP
standard preprocessor	actual filename: ASM21
core assembler	actual filename: ASM2

Different combinations of the modules can be run using the Assembler switches detailed below. Invocation of the Assembler with no switches runs the standard preprocessor and core assembler only.

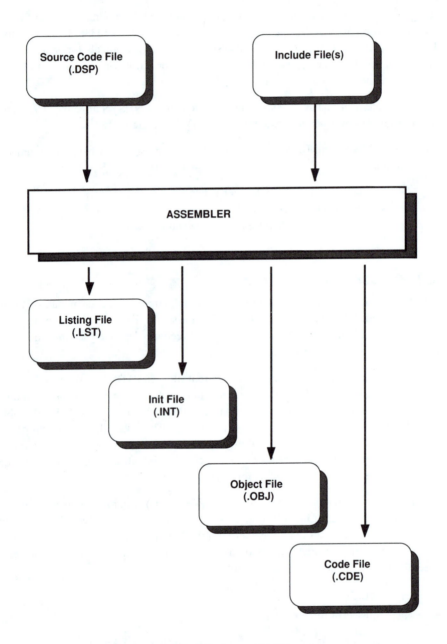

Figure 3-3: Assembler I/O

Running The Assembler

The Assembler is invoked from the host system by entering:

```
ASM21 filename[.ext] [-switch ...]
```

Filename[.ext] is the source code file. The filename extension is optional and defaults to .DSP. Other data and source code files are included in the assembly process using the directives .INIT and .INCLUDE (described later in this chapter).

The Assembler switches are not case-sensitive, and multiple switches must be separated by spaces. The Assembler switches are listed below in Table 3-3; some require arguments as shown. This list can be displayed on the monitor by invoking the Assembler with no filename or switches: ASM21.

Switch	Result
-cp	Runs C language preprocessor
-p	Runs standard preprocessor without core assembler
-d*variable*[=value]	Define *variable* for C preprocessor
-l	Creates .LST file
-m [number]	Macros expanded in .LST file, to depth of [number]
-i [number]	INCLUDE files expanded in .LST file, to depth of [number]
-s	No semantics checking
-c	Makes the Assembler case-sensitive

Table 3-3: Assembler Switches ————————————————————————————

A complete description of these switches is given in *ADSP-2101 Cross-Software Manual* [2].

Program Structure

The basic unit of an ADSP-2101 program is the module. Modules are defined by:

```
.MODULE[/qualifiers]   module_name;

STATEMENT;              (may be any of  • [label:] instruction
...                                     • directive
...                                     • macro invocation)

.ENDMOD;
```

Each element of the source module must end with a semicolon. Statements can be either an instruction, assembler directive, or macro call. Giving an instruction a label is optional. The .MODULE and .ENDMOD directives are defined in the section "Assembler Directives." Symbol names in the source code module must be unique. Assembler-reserved symbols may not be used as identifiers. Because the Assembler is not case sensitive, both upper and lower case keywords are reserved. Table 3-4 lists the assembler keywords. Some of those listed correspond to ADSP-2101 features which are not visible to users. Avoid them because their use may cause errors. Individual lines in each module must be no more than 200 characters in length.

ABS	DM	INCLUDE	MR0	RTS
AC	DO	INIT	MR1	RX0
AF	EMODE	JUMP	MR2	RX1
ALT_REG	ENA	L0	MSTAT	SAT
AND	ENDMACRO	L1	MV	SB
AR	ENDMOD	L2	MX0	SEG
AR_SAT	ENTRY	L3	MX1	SEGMENT
ASHIFT	EQ	L4	MY0	SET
ASTAT	EXP	L5	MY1	SHIFT
AUX	EXPADJ	L6	NAME	SI
AV	EXTERNAL	L7	NE	SR
AV_LATCH	FOREVER	LE	NEG	SR0
AX0	FLAG_IN	LOCAL	NEWPAGE	SR1
AX1	FLAG_OUT	LOOP	NOP	SS
AY0	GE	LSHIFT	NORM	SSTAT
AY1	GLOBAL	LT	NOT	STATIC
BIT_REV	GT	M0	OR	STS
BM	I0	M1	PASS	SU
BY	I1	M2	PC	TEST
C	I2	M3	PM	TIMER
CACHE	I3	M4	POP	TOGGL
CALL	I4	M5	PORT	TOPOFPCSTAC
CE	I5	M6	POS	TRA
CIRC	I6	M7	PRI	TRUE
CLR	I7	MACRO	PUSH	TX1
CLEAR	ICTRL	MF	RAM	TX0
CNTR	IDLE	M_MODE	REGBANK	UNTIL
CONST	IF	GO_MODE	RESET	US
DIS	IFC	MODIFY	RND	UU
DIVS	IMASK	MODULE	ROM	VAR
DIVQ		MR	RTI	XOR

Table 3-4: Asembler-Reserved Symbols/Keywords ————————————————

Assembler Directives

Assembler directives are instructions that control the assembly process. They do not produce opcodes. In the source file, an assembler directive statement starts with a period and ends with a semicolon. An assembler directive may take modifiers and arguments, as specified in each of the following sections. Some assembler directives which may take modifiers and arguments are briefly described below.

.MODULE Directive

The .MODULE directive defines the start of an assembly module and is the first statement. It has the form:

.MODULE[/qualifier…] *module_name*;

Qualifiers consist of any of the following:

RAM or ROM
ABS = absolute start address
BOOT = 0, 1, 2, 3, 4, 5, 6, or 7
SEG = memory segment name defined in System Builder

The module qualifiers determine the location of the module in memory. Memory type can be specified as RAM or ROM, followed by the start address and/or a physical segment in memory defined in the System Builder. (The start address is a constant.)

The example that follows defines the module *main_routine,* which is located at execution-time in RAM at address 0 (on-chip). The code is stored on boot page 0.

```
.MODULE/RAM/ABS=0/BOOT=0 main_routine;
```

The next example defines the module *filter_routine*, located in a memory segment named *fir* (as defined in a System Builder output .SYS file), which is specified as ROM.

```
.MODULE/ROM/SEG=fir filter_routine;
```

.ENDMOD Directive

The .ENDMOD directive is the last statement in a source file. The assembly process terminates when the Assembler reads the .ENDMOD directive. It has the form:

.ENDMOD;

.VAR Directive

The .VAR directive declares data buffers. All buffers must be declared with this directive prior to any use or reference to them. It has the form:

.VAR[/qualifier…] *buffer_name*[**length**],…;

One .VAR directive can have an unlimited number of declarations, each separated by commas, up to the maximum number of characters that can be processed. Specification of length is optional, with default to one (a single word variable).

Qualifiers consist of any of the following:

PM or DM
RAM or ROM
CIRC
ABS = absolute address
SEG = memory segment name defined in System Builder
STATIC

The following is an example of a variable declaration:

```
.VAR/DM/RAM/ABS=0x10F  seed;
```

This statement declares a one word variable called *seed* in data memory RAM, at hexadecimal address 10F.

The following is an example of one circular buffer of length five (three bits required to represent), which would be located by the Linker at an address that is a multiple of eight (has three LSBs equal to zero):

```
.VAR/CIRC aa[5];
```

.INIT Directive

The .INIT directive initializes a declared variable or all or part of a data buffer (in either DM or PM). The buffer is initialized with the value(s) listed or those contained in an external buffer. This directive has the form:

.INIT *buffer_name*:   constant or expression,...,
                          ^*other_buffer*[offset] or %*other_buffer*[offset],...,
                          <filename>;

Any combination of the three forms of initialization values shown above may be used, separated by commas. This directive recognizes the "pointer to" (^) and "length of" (%) operators.

In the following example, variable *seed* is initialized to a constant hex value.

```
.INIT seed: 0x3FFF;
```

In the second example, a variable *lookup_table* is set to point to the base address of buffer *sin*.

```
.INIT lookup_table: ^sin;
```

.CONST Directive

The .CONST directive declares symbolic constants. Symbolic constants can be used in place of numerical constants. The .CONST directive has the form:

.CONST *const_name*=**const or expression,...;**

One .CONST directive can have an unlimited number of assignment statements, each separated by commas, up to the maximum number of characters that can be processed. The following example defines two constants, equal to the numeric values shown.

```
.CONST taps=15, taps_less_one=14;
```

.PORT Directive

The .PORT directive declares a memory-mapped I/O port in data or program memory. The argument for this directive is a symbolic port name. The name must be the name of a port declared in the Architecture Description file. The .PORT directive has the form:

.PORT *port_name*;

The following example identifies the port *ad_sample* which has been previously declared as a specific memory location in the System Builder:

```
.PORT ad_sample;
```

.INCLUDE Directive

The .INCLUDE directive is used to include another source file in the file being assembled. The Assembler processes the included file as if it were part of the original source file. The .INCLUDE directive has the form:

.INCLUDE *<filename>*;

Source files specified by the .INCLUDE directive can have .INCLUDE statements within them (nesting of include files is limited only by memory). The .INCLUDE directive supports modular programming. For example, in many cases it is useful to develop a library of subroutines or macros which are shared between different programs. Rather than rewriting these routines for each program, you can incorporate a macro library into the source code file using the .INCLUDE directive. In the following example, file *macro_lib* is included while assembling.

```
.INCLUDE <macro_lib>;
```

Here the use of angle brackets is required.

.EXTERNAL Directive

The .EXTERNAL directive assigns the EXTERNAL attribute to variables, ports, and program memory labels declared in other assembly modules. Those symbols in other modules can only be referenced if they are assigned the external attribute in the referencing module and the global or entry attribute in the module where they are actually declared. It has the form:

.EXTERNAL *external_symbol,…;*

The following is an example.

```
.EXTERNAL fir_start;      {entry label in different module}
```

.GLOBAL Directive

The .GLOBAL directive assigns the GLOBAL attribute to variables, buffers and ports. Only such identifiers declared (with .VAR or .PORT) as global may be referenced in other modules. It has the form:

.GLOBAL *internal_symbol,…;*

A variable, buffer, or port that is declared within a module can be referenced only by that module unless explicitly specified as global. For program labels which are intended to be referenced in other modules, the .ENTRY directive rather than the .GLOBAL directive should be used. Example:

```
.GLOBAL seed;
```

Other modules are able to refer to global identifiers by declaring those symbols as EXTERNAL.

.ENTRY Directive

The .ENTRY directive assigns the ENTRY attribute to program labels. This makes the label visible to other modules for use in subroutine calls or inter-module jumps. This directive has the form:

.ENTRY *program_label,…;*

In the following example, the label *fir_start* is visible outside the current module.

```
.ENTRY fir_start;
```

Program Example

Listings 3-2 through 3-4 illustrate a sample source code program, an interrupt service subroutine, and an include file for the ADSP-2101. In this example the module *main_routine* is the main program and *fir_routine* is the subroutine. These modules are linked together to form a complete program.

There are six possible interrupt sources for the processor plus the restart vector at address 0. Each has four locations associated with it. As described in Chapter 1, the first 28 addresses in program memory contain the restart and interrupt vectors (0x0000 - 0x001B). The 29th PM address (0x001C) holds the first program instruction. Since *main_routine* is declared at absolute address zero, the first 28 instructions are placed in the interrupt vector locations. Because this example uses only the restart (0x0000) vector and SPORT0 Receive (0x000C) interrupt, the remaining instructions are simply returns (RTI).

The .VAR directive defines two circular buffers in on-chip memory: the one in data memory RAM is used to hold a delay line of samples and the one in program memory RAM is used to store coefficients for the filter. *Data_buffer* and *coefficient* are declared as GLOBAL buffers in the *main_routine*, while the *fir_routine* declares them as EXTERNAL. The address label, *fir_start*, is declared as ENTRY in *fir_routine* and can be referenced by *main_routine*, which declares it as EXTERNAL.

This sample program implements a FIR filter routine and has several features worth noting. After declaring the include file and memory buffers and performing initialization, *main_routine* jumps to location *restarter*. Here the data and coefficient buffers are cleared and the data memory-mapped control registers of the ADSP-2101 are set up. The functions selected include SPORT0 timing specification, μ-law companding, and 8-bit data words. SPORT0 interrupt is then enabled and the processor loops on the IDLE instruction until the interrupt from SPORT0 is received. The filter is thus interrupt-driven. When the interrupt occurs, the program control shifts to the subroutine by jumping to location *fir_start*.

All further activity takes place in the interrupt service routine, Listing 3-3. After the return from interrupt, execution resumes at the WAIT loop.

```
{ADSP-2101  FIR Filter program
Serial port 0 used for I/O
Internally generated serial clock
12.288 MHz clock rate gives 8000 Hz sampling rate}

.MODULE/RAM/ABS=0          main_routine; { program loaded from BOOT EPROM, }
.INCLUDE <const.h>;                      { MMAP=0 }
.VAR/DM/RAM/CIRC data_buffer[taps];    {data values}
.VAR/PM/RAM/CIRC coefficient[taps];
.GLOBAL data_buffer, coefficient;
.EXTERNAL fir_start;
.INIT coefficient: <coeff.dat>;        { initialize coeffs from external }
                                       { file }

                {code starts here}
                {load interrupt vector addresses}

                JUMP restarter; nop; nop; nop; {restart interrupt}
                RTI; nop; nop; nop;            {sampling interrupt IRQ2}
                RTI; nop; nop; nop;            {SPORT0 transmit int}
                JUMP fir_start; nop; nop; nop; {SPORT0 receive int}
                RTI; nop; nop; nop;            {SPORT1 transmit int}
                RTI; nop; nop; nop;            {SPORT1 receive int}
                RTI; nop; nop; nop;            {TIMER interrupt}

                {initializations}

Restarter:      L0 = %data_buffer;             {setup circular buffer length}
                L4 = %coefficient;             {setup circular buffer length}
                M0 = 1;                        {modify=1 for increment}
                M4 = 1;                        {through buffers}
                I0 = ^data_buffer;             {point to data start}
                I4 = ^coefficient;             {point to coeff start}
                CNTR = %data_buffer;           {setup loop counter}
                DO clear_buffer UNTIL CE;
clear_buffer:   DM(I0,M0)=0;                   {clear data buffer}
```

```
                    AX0=0x1000;
                    DM(sys_cont_reg)=AX0;              {SPORT0 enabled }
                    AX0=0;
                    DM(dm_wait_reg)=AX0;              {All DM wait states 0}
                    AX0=0x6B27;
                    DM(sport0_cont_reg)=AX0;          {Sport0 control register}
                    AX0=2;
                    DM(sport0_sclkdiv)=AX0;           {Generate 2.048MHz serial clk}
                    AX0=255;
                    DM(sport0_rfsdiv)=AX0;            {Divide by 256 for 8 KHz rate}

                    ICNTL = 0x07;                     {Enable edge sensitive int}
                    IMASK = 0x0018;                   {enable SPORT0 interrupt only}

WAIT:               IDLE;                             {wait for interrupt}
                    JUMP WAIT;

.ENDMOD;
```

──────────────── Listing 3-2: Main Routine Example ────────────────

```
.MODULE/RAM fir_routine;                   {relocatable interrupt service }
                                           { routine module}
.INCLUDE <const.h>;                        {include constant declarations}
.ENTRY fir_start;                          {make label visible outside }
                                           { module}
.EXTERNAL data_buffer, coefficient;        {make global buffers visible to }
                                           { module}


                    {code}

FIR_START:          CNTR = taps-1;           {N-1 passes within DO loop}
                    SI = RX0;                {read from SPORT0}
                    DM(I0,M0) = SI;          {transfer data to buffer}
                    MR=0, MY0=PM(I4,M4), MX0=DM(I0,M0);
                                             {set up multiplier for loop}

                    DO convolution UNTIL CE; {CE = counter expired}
convolution:        MR=MR+MX0*MY0(SS), MY0=PM(I4,M4), MX0=DM(I0,M0);
                                             {MAC these, fetch next}

                    MR=MR+MX0*MY0(RND);      {Nth pass with rounding}
                    IF MV SAT MR;            {saturate if overflowed}
                    TX0 = MR1;               {write to sport 0 transmit}
                    RTI;                     {return from interrupt}
.ENDMOD;
```

──────────────── Listing 3-3: Interrupt Routine Example ────────────────

```
.const sys_cont_reg=0x3FFF;
.const dm_wait_reg=0x3FFE;
.const sport0_cont_reg=0x3FF6;
.const sport0_sclkdiv=0x3FF5;
.const sport0_rfsdiv=0x3FF4;
.const taps=15;
```

──────────────── Listing 3-4: Include File, Constant Initialization ────────────────

In summary, the Assembler:

- supports high level constructs,
- encourages modular code development, and
- provides a full range of diagnostics

3.2.3 The Linker

The ADSP-2101 Linker generates a complete executable program by linking together program modules which were assembled separately. It can search libraries, which are simply subdirectories, for subroutines to link. The output of the Linker is used by the Emulator, Simulator and PROM Splitter. Figure 3-4, on the following page, shows the files read and created by the Linker.

As shown in the previous section, the Assembler processes each source code module separately, producing an Object file (.OBJ), a Code file (.CDE) and an Initialization file (.INT), which contains information on the assembled code, source level declarations and initialization information. Initialization data files (.DAT) are created separately. Changes in initialization data only require relinking.

The Assembler output files (one set for each module to be linked), together with initialization data files and the Architecture Description file are used by the Linker. The Linker expects to find an Architecture Description file with the default name 210x.ACH unless the user alters this name with a switch; the files to be linked must be specified in the invocation command or located in libraries to be searched.

The Linker creates one complete executable code file by resolving external references and assigning addresses to relocatable code and data spaces.

The Linker can generate three files. The Memory Image file (.EXE) is always created, and contains the actual program memory, data memory, and boot memory images after the linkage. This file is used by the Simulator and Emulator, and is also passed to the PROM Splitter to prepare a data file for a PROM burner. It has the default name 210x.EXE which can also be changed with a switch.

The optional map listing file (.MAP) assists the user in interpreting the result of the linkage. This file is discussed in more detail later in this section. The optional debug symbol table file (.SYM) lists all symbols encountered by the Linker, their absolute values and their scope of reference. This file is used by the Simulator and Emulator.

The Linker can link together an unlimited number of modules and initialization data files. The initialization data files (.DAT) are not explicitly named in the invocation line because they are specified (with the .INIT directive) in the source code files. The data files are incorporated by the Linker. When changes are made in the data files, simply relink the modules to incorporate the new data file.

Running the Linker

To invoke the Linker from the host system, the command form is:

```
LD21 file1 [file2 .] [-switch .]
```
or
```
LD21 -i file_all [-switch .]
```

The -i switch causes the Linker to read the file *file_all* for a list of files to link. The file containing the list of files to link must be a simple text file with one pathname/file per line.

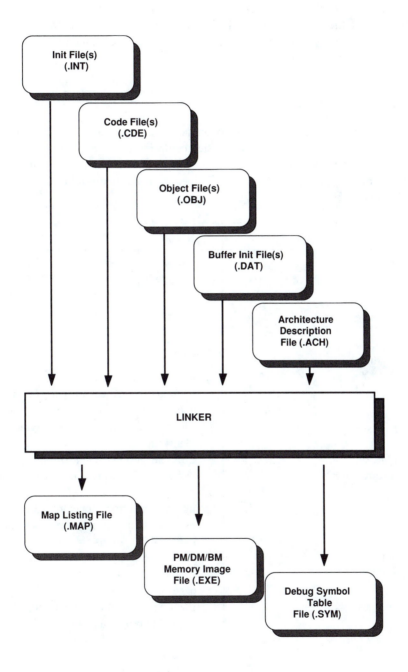

Figure 3-4: Linker I/O

In the first form, the user explicitly names all the files to be linked (separated by spaces). In both forms, the filename(s) must identify the Assembler output files (.CDE, .OBJ and .INT) without any extension. Modules to link are searched for in the current directory or in the pathname specified in the command line.

Linker Switches

The switch component of the invocation command can have any of the Linker switches (separated by spaces). The Linker switches are listed below in Table 3-5; some require arguments as shown. This list can be displayed on monitor by invoking the Assembler with no filename or switches: LD21. A complete description of these switches is given in *ADSP-2101 Cross-Software Manual* [2].

Switch	Result
-a *archname*	Use *archname.ACH* Architecture Description file instead of default 210x.ACH
-c	Linker creates "top of RAM" symbol to locate the stack; this symbol is used by programs generated with the ADSP-2101 C Compiler (See Chapter 7)
-dryrun	Linker does not generate an .EXE file; quick test to check for link errors
-e *target*	Output files named *target.EXE*, instead of default 210x.EXE
-g	Linker generates a debugger symbol table, .SYM file
-i *file_all*	Links all files listed in text file *file_all*
-lib *directory*; ...	Directories listed are added to those found in ADIL environment setting for locating libraries; multiple directories are separated by commas in Unix systems or by semicolons in PC-DOS systems
-old	Not used (ADSP-2100 feature)
-p	Library subroutines are assigned to the boot pages that call them
-pmstack	Used with -c switch; moves "top of RAM" symbol to program memory
-s *stack_size*	Used with -c switch; specify a maximum size for stack
-x	Linker generates a .MAP file

Table 3-5: Linker Switches

MAP Listing File

The Map Listing file is generated to help the user interpret the Linker result. The file provides information on:

- Symbols
  A cross-reference listing of all symbols encountered, arranged by module. For each module a list is shown of the symbols referenced in that module, with the following information for each symbol: its absolute address, its length, the type of symbol (module, variable, or label) , and the type of memory (PM, DM, or BM).

- Memory segments
  A map of the physical memory segments declared for the system with the absolute address, length, and attributes of each. The information here reflects the content of the Architecture Description file.

- Boot memory & Run-time program memory
  An address map of modules and data structures on each boot page, and the corresponding map of booted code in internal program memory ("bootable run-time program memory"). Information on PROM byte addresses and boot PROM sizes required is also provided.

- Fixed vs. Dynamic memory
  Maps of fixed program memory, dynamic data memory, and fixed data memory. These maps include address, length, and attribute specifications.

- Error messages
  Linker error messages.

- Libraries
  A list of libraries searched and used.

A sample Map Listing file is shown in Listing 3-5.

```
ADSP-210x Linker, version 2.02, copyright Analog Devices, Inc.
210x (210x.exe) mapped according to EZLAB_SYS (ezlab1.ach)

xref for module: MAIN_ROUTINE
    MAIN_ROUTINE              pm 0000 [0033]    module(global)
    DATA_BUFFER              dm 3800 [000F]    variable(global)
    COEFFICIENT              pm 0040 [000F]    variable(global)
    RESTARTER                pm 001C           label
    CLEAR_BUFFER             pm 0024           label
    WAIT                     pm 0031           label
    FIR_START                   0033 [0000]    extern(FIR_ROUTINE)

xref for module: FIR_ROUTINE
    FIR_ROUTINE              pm 0033 [000A]    module(global)
    FIR_START               pm 0033           label
```

```
        CONVOLUTION                      pm 0038              label
        COEFFICIENT                         0040 [000F]       extern(MAIN_ROUTINE)
        DATA_BUFFER                         3800 [000F]       extern(MAIN_ROUTINE)

210x memory per EZLAB_SYS (ezlab1.ach)

        0000 - 07FF [   2048.] pm ram data/code INT_PM
        3800 - 3BFF [   1024.] dm ram data INT_DM

boot memory and bootable run time program memory map:

fixed program memory map:

        0000 - 0032 [    51.] pm ram        module   MAIN_ROUTINE of MAIN_ROUTINE
        0033 - 003C [    10.] pm ram        module   FIR_ROUTINE of FIR_ROUTINE
        0040 - 004E [    15.] pm ram circ variable COEFFICIENT of MAIN_ROUTINE

        fixed program memory rom:       0.
        fixed program memory ram:      76.

dynamic data memory map:
        fixed data memory map:
        3800 - 380E [     15.] dm ram circ variable DATA_BUFFER of MAIN_ROU-
TINE

fixed data memory rom:       0.

        fixed data memory ram:       15.
```

──────────── Listing 3-5: MAP Listing File ─────────────────────

In summary, the Linker:

- supports multi-module linking, and
- maps the Assembler output to the system architecture.

3.2.4 The Simulator

The ADSP-2101 Simulator is an interactive window-oriented software tool for instruction level simulation and debugging of a user program. The Simulator configures itself according to the user's target system architecture as defined in the Architecture Description file (.ACH). This allows it to flag illegal operations such as reading from non-existent memory. Using the symbol table created by the Linker, the Simulator is able to provide a fully symbolic environment for simulation and debugging.

Briefly, the Simulator provides the following functions:

- Instruction level simulation of booting and execution
- Simulation of ports and SPORTs using host data files
- Simulation of internal and external interrupts
- Complete assembly and disassembly of the ADSP-2101 instruction set
- Multiple break conditions including break at address, break on condition, break on expression and break on address ranges
- Full view of all processor registers and the ability to directly change any register's contents interactively

The simulator uses a variety of files as shown in Figure 3-5. The inputs are the architecture description file (.ACH), the Program/Data Memory image (.EXE) and debug symbol table (.SYM) generated by the Linker, simulated input data buffers (.DAT) downloaded from the host computer, and simulated inputs. Simulator outputs are processed signals uploaded to the host and simulated outputs to I/O ports. The primary Simulator diagnostics are displays of program execution, register contents and chip status, stacks, PM/DM contents, etc. Upon first booting the Simulator, the user sees the command window display as shown in Figure 3-6. From this window, the user can open, configure and use all other features of the simulator. Typing Ctrl-W (control-w) displays a menu of window commands including, for example, OPEN, which in turn displays a submenu of windows to be opened.

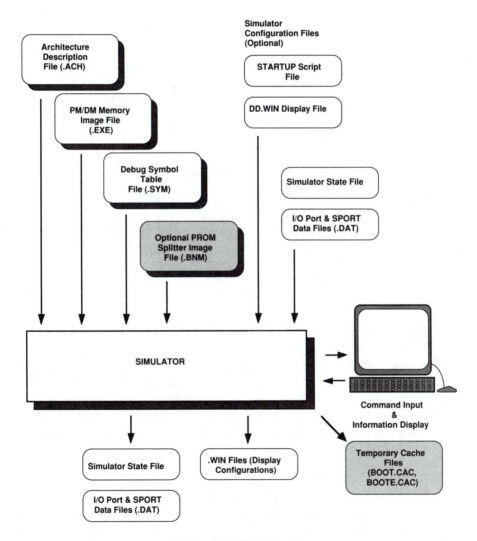

Figure 3-5: Simulator I/O

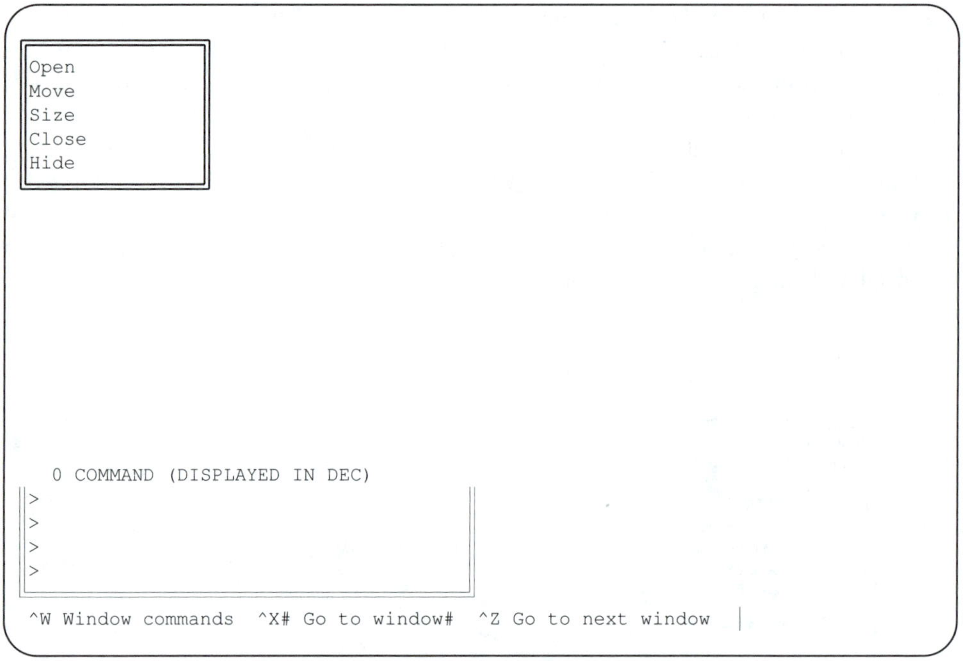

```
Open
Move
Size
Close
Hide




     0 COMMAND (DISPLAYED IN DEC)
   >
   >
   >
   >
 ^W Window commands  ^X# Go to window#  ^Z Go to next window   |
```

Figure 3-6: Simulator Command Window

The user can customize the contents and layout of many windows, the arrangement of multiple windows on the screen and the command strings used to invoke various Simulator functions. All customized settings can be stored in an external file and invoked automatically upon startup. A complete illustrative example of screen configuration and window manipulation is given in Chapter 4 section 4.3.1. For details, consult the *Cross-Software Manual* [2].

Invoking The Simulator

The Simulator invocation command is:

sim2101 [-a *archname*] [-w *window*] [-s *script*]

The -a switch uses a unique .ACH Architecture Description file that is used to link programs. The Simulator configures itself according to this target architecture. A default filename 210x.ACH is assumed. The optional -w switch identifies a .WIN file containing a stored windows configuration which is loaded as the initial display when the simulator is first booted. The optional -s switch identifies a file containing Simulator commands to be executed automatically upon startup.

Simulator Function Overview

The Simulator generally provides multiple methods for achieving a given result. For example, there are two different methods for setting breakpoints in program memory. Consequently, it makes sense to think of the Simulator's *functions* rather than *command structure*. The functional capabilities of the Simulator are grouped into these broad classes:

- Interface management functions

  These functions include the opening and closing of windows, changing the size and position of windows, and changing the appearance of a window (removing or adding items to the window and rearranging the items displayed within the window's space). Additional functions described under this heading include navigating from window to window. Saving specific window configurations is possible and is described in the next chapter. Aliasing commands is another aspect of interface management.

- Set-up functions

  These include loading the program to be simulated, opening I/O ports and associating data files with I/O ports and SPORTs for the purposes of simulating input and output data streams. Also included is the configuring of simulated interrupts.

- Register inspection & change functions

  These functions allow you to view the contents of all the registers in the processor and, in most cases, to change their contents directly if desired. Several windows are dedicated to register displays.

- Memory inspection & change functions

  These functions include simple display of the various memory spaces (as either data or code), saving the contents of memory to files for later analysis and plotting the contents of data memory.

- Simulator control & debugging functions

  Control functions include starting and stopping the execution of your program and resetting the simulated processor. Debugging functions include setting breakpoints, break conditions and watchpoints. The Simulator supports a wide variety of break expressions for debugging purposes.

A more detailed discussion including examples is given in Chapter 4. In summary, the Simulator:

- provides an interactive and user-friendly interface,
- supports full symbolic assembly and disassembly,
- simulates hardware configuration,
- simulates interrupt and I/O handling,
- flags illegal operations, and
- displays the internal operations and status of the processor.

3.2.5 The PROM Splitter

The ADSP-2101 PROM Splitter extracts the address information and the contents of the ROM portion of the memory image file (.EXE) and formats the extracted images for uploading to PROM burners.

The PROM Splitter creates output files for program, data, and boot memory. Three usable files are created for PM to organize the PROMs in word addresses corresponding to three-byte instructions. Two usable files are created for DM to organize any data PROMs in terms of two-byte data words. One usable file is created for BM, which is physically byte-wide although organized internally in vertical groups of four bytes per word address. Both program and data memory can also be optionally output as a single stream of bytes for vertical rather than horizontal grouping of words in the PROMs.

Since the PROM Splitter is more appropriate for commercial use, we will not discuss it any further. For more information, consult the *Cross-Software Manual* [2].

3.3 HARDWARE DEVELOPMENT TOOLS

The hardware development tools for the ADSP-2101 consists of emulators in various degrees of capabilities from high end to low end. A full function emulator with access to various parts of the chip is manufactured by Microtek. At the intermediate level is an in-circuit emulator called EZ-ICE™ with moderate control over the processor. At the low end is a low cost evaluation system called EZ-LAB™ which can be used as a demonstration as well as a target system. In this section we briefly describe the EZ-Tools consisting of the EZ-ICE and EZ-LAB systems since they are appropriate in an educational environment.

3.3.1 The In-Circuit Emulator - EZ-ICE™

The ADSP-2101 EZ-ICE is a compact, easy-to-use in-circuit emulator for debugging code and testing ADSP-2101 based systems. It is a 3.3" x 3.3" in-circuit probe board containing a 121-pin emulator version of the ADSP-2101. A 68-pin PGA footprint protrudes from the bottom of the board. These pins are inserted into a socket in a target system. EZ-ICE requires a +5 V dc power supply capable of supplying 1 A of current.

EZ-ICE can be run at full speed (12.5 MIPS). There is no degradation of ADSP-2101 performance or signal timing other than $\overline{BR}$, $\overline{BG}$ and $\overline{RESET}$. The user can select via a jumper either the target system clock or the EZ-ICE clock. The oscillator is socketed to allow the use of other oscillator devices to achieve different clock speeds.

For display and input, EZ-ICE requires either a VT100 as the terminal device or a personal computer (PC) running a terminal emulation program. The program should be capable of emulating a VT100-type terminal and allow the transfer of ASCII files between the PC and EZ-ICE. EZ-ICE is connected to the PC via an RS-232 cable. It automatically adjusts its baud rate to match the host PC's baud rate. A baud rate of 9600 or 19200 is recommended.

The user has the option of running ADSP-2101 programs from target system memory, emulator overlay memory or a combination of both. The 8 K by 24-bit overlay program/data memory option is jumper selectable.

The monitor firmware in EZ-ICE controls all emulator functions. The user interface is simple. There are no commands to remember; everything is menu- or cursor-controlled. In addition, EZ-ICE firmware intercepts illegal user inputs, making debugging work easier.

Control and debug features include single-step capabilities with or without register displays and a multiple breakpoint capability (with up to 16 breakpoints individually set).

At power-up, the host processor is automatically reset and a diagnostic check is performed to ensure that both host memory and EZ-ICE are functional. A report of any failures found is automatically displayed.

EZ-ICE Features

The following is a summary of the EZ-ICE features:
- Stand alone operation via RS-232 to a PC
- In-Circuit self-emulation plugs directly into a target board
- Full speed emulation
- Single step capability
- Multiple break points
- 8K overlay memory

EZ-ICE vs. Full Featured Emulator

EZ-ICE is a basic, easy-to-use emulator that does not have some of the advanced features of the ADSP-2101 full featured emulator. It has the following limitations:

- No trace capability
- No hardware event triggers
- Breakpoints for program memory instructions only
- No data memory breakpoint (watchpoints)
- Can not monitor the state of all ADSP-2101 pins
- Can not plot memory contents
- No symbolic debug and online assembly is provided by emulator firmware, only disassembly on program memory reads
- No session record is kept by emulator firmware
- Can not use breakpoints with internal ADSP-2101 memory while in GO mode.

EZ-ICE Functions

The EZ-ICE emulator uses Simulator-compatible memory image files. However EZ-ICE can run stand alone unlike the Simulator which runs only on the host. The emulator has the capability to set software breakpoints, upload and download PM and DM to and from the host, inspect and change all internal registers, halt execution and single-step, and switch memory banks between emulator and target.

The hardware component layout is shown in Figure 3-7.

The basic command menu of the emulator firmware is shown in Figure 3-8. It is discussed in more detail in Chapter 4.

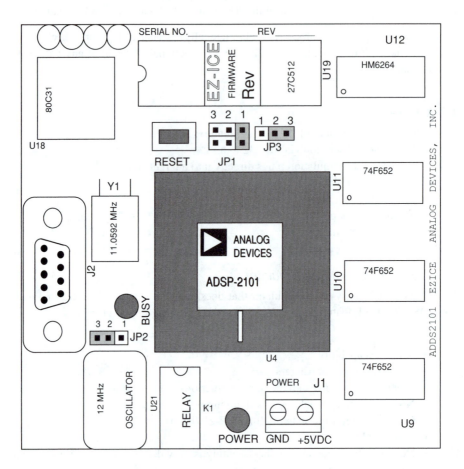

Figure 3-7: EZ-ICE Component Lay Out

3.3.2 The Evaluation System - EZ-LAB™

The ADSP-2101 EZ-LAB demonstration and evaluation board is a complete system on a 4 1/2"
by 6" board. It allows the user to test coded applications in real time. No host or PC is needed
to operate EZ-LAB. At reset, the ADSP-2101 on EZ-LAB boots code and program memory
data into its internal program memory from a 64 K x 8-bit EPROM. It then executes the code.
EZ-LAB is capable of stand-alone operation. It requires a +5 V dc power supply capable of
supplying 500 mA and a ±12 V dc power supply capable of supplying 200 mA with a common
power return.

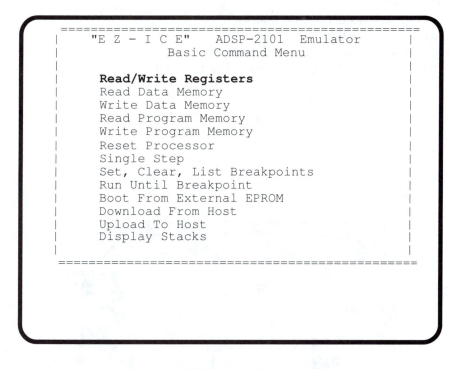

```
===================================================
|     "E Z - I C E"    ADSP-2101   Emulator        |
|               Basic Command Menu                  |
|                                                   |
|     Read/Write Registers                          |
|     Read Data Memory                              |
|     Write Data Memory                             |
|     Read Program Memory                           |
|     Write Program Memory                          |
|     Reset Processor                               |
|     Single Step                                   |
|     Set, Clear, List Breakpoints                  |
|     Run Until Breakpoint                          |
|     Boot From External EPROM                      |
|     Download From Host                            |
|     Upload To Host                                |
|     Display Stacks                                |
|                                                   |
===================================================
```

Figure 3-8: EZ-ICE Basic Command Menu

A codec is attached to one of the serial ports. The other serial port can be configured for interrupts and flags by changing on-board jumpers. The input signal to the codec can be a microphone, signal generator or any other high impedance source, and the resulting output signal can drive a small speaker.

EZ-LAB has four DAC outputs to connect to an oscilloscope for display. In addition, there is an expansion connector and a serial port connector. The connectors allow access to the ADSP-2101's serial ports, external address bus, external data bus, control signals, and interrupt lines.

EZ-LAB provides manual control of several functions: pushbuttons assert the ADSP-2101's IRQ2 interrupt and FLAG IN pins and an on-board hardware RESET switch reset EZ-LAB.

As a hardware development system for this book, we will combine EZ-LAB and EZ-ICE to form a high speed DSP workstation with an interactive, window-based debugging interface. This can be done by simply removing the ADSP-2101 device from the EZ-LAB board and plugging in EZ-ICE. This combination will allow us to prototype and evaluate our experiments and application with virtually no initial time investment in hardware design.

EZ-LAB Features

The following a summary of EZ-LAB' features:

- ADSP-2101 12.5 MHz microcomputer,
- programs can be loaded into the internal program memory of the ADSP-2101 from a standard, low-cost EPROM; no external memory board is needed,
- manual control of several EZ-LAB functions,
- codec connected to SPORT0,
- the input signal to the codec can be a microphone, or signal generator, and the resulting output signal can drive a small speaker,
- a small size circuit board.

The EZ-LAB board layout is shown in Figure 3-9.

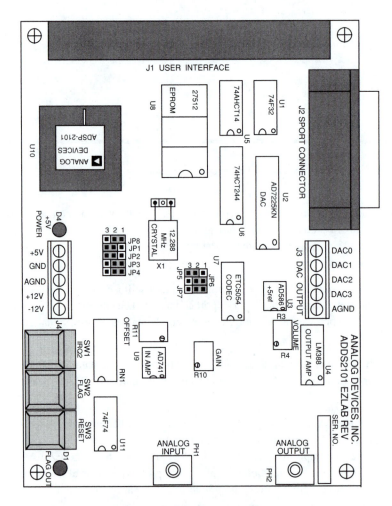

Figure 3-9: EZ-LAB Board LAyout

3.4 HOST PC REQUIREMENTS

The software development tools described above are run on a host computer. Although this host can be any computer platform, it is convenient to use personal computer in an educational environment of a digital signal processing laboratory.

The ADSP-2101 Cross-Software executes on a IBM-AT or compatible PC with a hard disk. Installation of the Cross-Software requires PC-DOS 3.0 or later, 640 KBytes Memory, and a color display system : CGA, EGA, or VGA type.

Although the emulator program executes from firmware on the board, it sends menus, the results of operations, and files to the host. Similarly it accepts instructions and files from the host. This is done through the serial port of the PC via a RS-232 cable. Therefore the host PC must run a communications program, such as Procomm, to imitate a VT100 type terminal. The communications program must be capable of setting the baud rate for the serial port at 9600 or 19200. The communications parameters are: no parity, 8 data bits, and 1 stop bit. The communications program should also be capable of transferring files between emulator and host using an ASCII format

Finally to create source program files, the PC must have a screen editor.

3.5 SUMMARY

In this chapter we described an important element in the environment of the ADSP-2101 based systems: the development system. Several software tools from system builder to PROM splitter were explained and two hardware tools were discussed. Requirements on a host PC were also described.

chapter 4

GETTING STARTED WITH
THE ADSP-2101

4.1 INTRODUCTION

In the first three chapters, we described hardware, software as well as development tools required to understand, program and integrate the ADSP-2101 microcomputer in a laboratory environment. Even though it is possible to comprehend this description after reading these chapters, it may be difficult, to use this enormous information effectively in any application. Each aspect of the system development must be carefully studied and practiced to gain insight and experience. This chapter is designed to provide such an experience. It provides hands-on training on important aspects of system development so that the students can be brought up to minimum speed in order to undertake more elaborate experiments and projects.

In Section 4.2, we begin with the System Builder tool which describes the target ADSP-2101 system. This is an important aspect in an overall design. An accurate and complete description of the target system avoids unwarranted access to memory or I/O locations and optimizes program coding. After providing a complete description of System Builder directives, we discuss three sample target systems and explain how to write their architecture description file. We complete this section with more target systems and exercises.

Section 4.3 deals with the Simulator which is a very important software tool. It allows us to perform instruction-level simulation. Its user interface is both interactive and symbolic. We give a sample session to acquaint the student of its various windows and customized screens and commands. A complete window command description is then given. The Simulator is also an excellent vehicle to learn the instruction set of the ADSP-2101. Therefore we provide an instruction set work-out involving various types of instructions.

The hardware emulator, the EZ-ICE, is studied in Section 4.4. It is the final link in the overall development cycle. In the section, we first provide a discussion on the proper power-up and power-down operational sequences for the emulator. A complete description of its firmware and window commands then follows.

4.2 SYSTEM BUILDER

The System Builder is a software tool which is used to describe the target system. The ADSP-2101 is located in this target system which supports it and might contain A/D and D/A converters, a ROM to hold program memory and several RAMs for proper execution and operation. It must be pointed out that this target system is not unique and that every product or application that is built around the ADSP-2101 microcomputer has its own target environment. Therefore the ADSP-2101 must be aware of its environment so that it will not access memory locations that are non-existent or interface with unconnected I/O devices. This is done through a System Specification File (.SYS) which is created using a text editor on a computer. The system specification file specifies the amount of RAM and ROM available, the allocation of program and data memory, memory-mapped I/O ports, and a host of other important information. The System Builder processes the .SYS file and generates an Architecture Description File (.ACH). This file is used by the Linker to resolve relocatable program and/or data memory. It is also used by the Simulator and the Emulator to setup a target environment.

In this section, we will examine some sample target systems and write their specification files. The specification file is essentially a program written using System Builder directives. Therefore we will begin with a detailed description of these directives. We will then execute the System Builder on the sample system's specification files to generate architecture description .ACH files. Finally we will provide some exercises on the System Builder.

4.2.1 System Builder Directives

This section describes each system builder directive and its syntax.

.SYSTEM Directive

The .SYSTEM directive must be the first statement in the System Specification source file. The identifier name given as its argument is the name of the system displayed in the Simulator. The .SYSTEM directive has the form:

.SYSTEM *system_name*;

.ENDSYS Directive

The .ENDSYS directive must be the last statement in the file. The System Builder processing terminates at the .ENDSYS directive statement. The .ENDSYS directive has the form:

.ENDSYS;

.ADSP2101 Directive

This directive identifies the processor. Its use is mandatory to clearly differentiate between ADSP-2100-based and ADSP-2101-based systems. If the directive is not present, the Cross-Software system assumes that the processor is an ADSP-2100.

.CONST Directive

The .CONST directive defines System Builder constants. Once you declare a constant, you may use it in place of its numeric value. This symbolic constant is recognized only by the System Builder, the definition is not carried over to the Assembler or Simulator. The .CONST directive has the form:

.CONST *constant_name* = constant or expression, ... ;

A single .CONST directive may declare one or several constants, separated by commas.

If you wish to define the value 15 for the term *taps*, for example, the directive would be as follows:

```
.CONST taps = 15;
```

.PORT Directive

The .PORT directive declares a memory-mapped parallel I/O port. Ports can be placed in either data or program memory, and must be declared in one or the other. The directive takes the absolute physical address of the I/O port as a modifier, and the symbolic name of the port as an argument. The .PORT directive has the form:

.PORT/qualifier ...   *port_name*;

There are two required qualifiers:

PM or DM   (in which memory space)
ABS=address   (absolute address (constant))

The port address is specified by a constant; *port_name* is an identifier. For example,

```
.PORT/DM/ABS=0x0400 ad_sample;
```

declares a port identified as *ad_sample* located at absolute data memory address 1024 (decimal). Assembler references to this same symbolic name are correctly interpreted by the Linker, using the .ACH file information.

.MMAP Directive

The .MMAP directive specifies the state of the MMAP pin on the ADSP-2101. It has the form .MMAP0 (MMAP pin held LO) or .MMAP1 (MMAP pin held HI). If .MMAP0 is used, boot loading takes place and on-chip program memory begins at address zero. If .MMAP1 is used, no boot loading takes place and on-chip program memory is mapped at the top of the program memory space. When this directive is omitted, the default is to .MMAP0.

.SEG Directive

The .SEG directive names a specific section of physical memory in the target system, and describes its attributes. In effect, the default memory map from the perspective of the System Builder is no memory at all. Until you declare and define a memory segment it does not exist. The .SEG directive has the form:

.SEG/qualifier ...   *seg_name[length]*;

The following qualifiers are mandatory:

PM or DM or BOOT=0, 1, 2, 3, 4, 5, 6, 7   (in which memory space)
RAM or ROM   (memory type)

While the following are optional:

ABS=address   (absolute start address (constant))
DATA or CODE or DATA/CODE   (what is stored in segment)

Seg_name is an identifier; *length*, which must be a constant or expression enclosed in brackets, is the number of words in the segment.

The .SEG directive declares three types of memory segments: program memory (PM), data memory (DM) and boot memory (BOOT). Qualifiers may specify the absolute start address of the segment, the physical memory type (RAM or ROM) and what is stored (DATA and/or CODE).

PM memory segments can be either CODE only, DATA only, or both CODE and DATA (defaults to CODE). For a PM segment that contains code and data, both modifiers must be used in the directive statement. The processor requires that any data access to PM must be made to sections with the DATA attribute. If a system requires that executable code be read or written by the processor as data, these sections should be declared with both CODE and DATA attributes.

DM memory segments must be DATA only. Therefore, the /DATA modifier can be omitted. An error is generated if a DM segment is assigned the CODE attribute.

BOOT memory segments may be either ROM- or RAM-type; in most systems, however, the boot memory chips are PROM and all BOOT segments are specified as ROM-type. Boot memory always defaults to both CODE and DATA; the CODE and DATA attributes are unnecessary. The BOOT modifier always specifies the page number, for example, BOOT=0. A system may have up to 8 boot pages, with page numbers from 0 to 7. Each page can hold up to 2K words of code and data. The System Builder knows how long a page can be and the possible boundaries for each page; it ignores the ABS modifier for boot pages. An individual declaration must be made for each boot page required.

Memory segments are assigned symbolic names. In the Assembler you may locate individual code modules and data objects (buffers and variables) in segments by name. The Assembler accepts the segment references; the Linker resolves them using the .ACH file.

The length of the segment is specified by the bracketed expression, as in *somedata[1024]*. The unit is always words, either 16-bit data or 24-bit instructions. This means that data memory segment size in bytes is 2x the word count, program memory size in bytes is 3x the word count and boot memory size is 4x the word count. The latter reflects the padding of boot memory with an extraneous byte per instruction in order to place the beginning of every instruction on an even byte boundary.

The example

```
.SEG/BOOT=0/ROM boot_mem[2048];
```

declares the boot segment, *boot_mem*, which is physical memory type ROM, residing in boot page zero (corresponding automatically to absolute address 0). The length of the segment is 2048 words corresponding to one page of boot memory.

The example

```
.CONST onchip_pm = 2048;
.SEG/PM/RAM/ABS=0/CODE/DATA  int_pm[onchip_pm];
```

declares a program memory segment called *int_pm*, which is memory type RAM at absolute location 0. This segment may hold both code and data. The length of the segment is 2048 words. This corresponds to the ADSP-2101 on-chip program memory space.

4.2.2 System Architecture Examples

Target systems are generally provided using a logic block diagram. Sometimes a text description is available. We will use both approaches to study target systems.

Example 4-1:

Consider a target system shown in Figure 4-1.

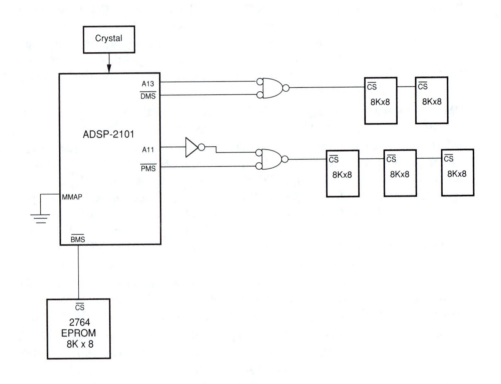

Figure 4-1: Example 4-1 System Architecture (Note: Data, Address, and Control signals are not shown.)

Observations:
- The ADSP-2101 microcomputer is in the system.
- The MMAP pin is held LO.
- 8K EPROM (8-bit) containing boot memory is connected. Since the program memory is 24-bit (3-byte) wide and since every 4th byte is not used (refer to Chapter 1), the boot memory length is 2K.
- 8K external data memory (16-bit) is selected when address line A13 is zero. Hence

the memory chip will be selected at any value below $2^{13} = 0x2000 = 8196$ beginning at absolute address 0 (refer to Data Memory Map in Chapter 1).
- 8K of external program memory (24-bit) is selected when address line A11 is one. Hence the memory chips will be selected beginning at absolute address 0x0800 or 2048 (refer to program memory map for MMAP=0 in Chapter 1).
- This example system does not have any I/O ports declared.

The System Specification File for describing the above system is shown below. Let us call it EXMPL_1.SYS. Note that program lines must end with a semicolon (;) and that comment fields are enclosed within braces, { }.

```
.SYSTEM                        EXAMPLE_1;      {System Name}
.ADSP2101;                                     {ADSP-2101 system}
.MMAP0;                                        {Enable Boot loading}
.SEG/BOOT=0/ROM                boot_mem[2048]; {Boot Segment}
.SEG/RAM/ABS=0x3800/DM/DATA    int_dm[1024];   {Internal Data Memory}
.SEG/RAM/ABS=0/DM/DATA         ext_dm[8192];   {External Data Memory}
.SEG/RAM/ABS=0/PM/CODE/DATA    int_pm[2048];   {Internal Program Memory}
.SEG/RAM/ABS=0x0800/PM/CODE/DATA ext_pm[8192]; {External Program Memory}
.ENDSYS;                                       {End of File}
```

Explanations:
- The first directive must be a .SYSTEM which also gives a name to the system.
- The second directive establishes an ADSP-2101 system.
- The third directive specifies that boot loading will take place upon reset and that on-chip program memory will begin at address zero.
- The fourth directive identifies a 2K-word space for one page of external boot memory.
- The next two directives declare data memory space. The first of these declares the 1K-word on-chip data memory which always begins at address 0x3800 or 14336 (decimal). The second one declares the 8K-word external memory beginning at address zero.
- The next two directives declare program memory space. The first of these declares the 2K-word on-chip program memory which begins at address zero since MMAP=0. The second one declares the 8K-word external memory beginning at address 0x0800 or 2048 (decimal).
- Finally, the last directive signals the end of the target system description file.

The architecture description file, EXMPL_1.ACH, can now be generated by executing System Builder on EXMPL_1.SYS.

BLD21 EXMPL_1

The system Builder responds with error messages, if any, or with a summary of the architecture created to the screen, which is shown below. Note that .ACH file is a binary file and as such should not be printed or listed.

```
boot memory
0000-07ff [0800] Rom code/data BOOT_MEM
program memory
0000-07ff [0800] ram data       INT_PM
0800-27ff [2000] ram code/data 24 required bits in memory width EXT_PM
data memory
0000-1fff [2000] ram data       EXT_DM
3800-3bff [0400] ram data       INT_DM
```

The System Builder alerts the user with a message "24 required bits in memory width EXT_PM" which is satisfied by the above system.

In the next example, we will include some I/O ports in the target system.

Example 4-2:

Figure 4-2 shows another target system in a block diagram form.

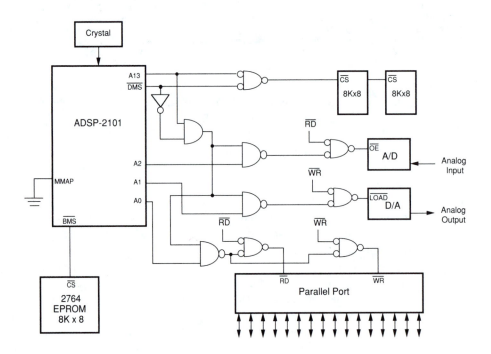

Figure 4-2: Example 4-2 System Architecture (Note: Data, Address, and Control signals are not shown.)

Observations:
- The ADSP-2101 microcomputer is in the system.
- The MMAP pin is held LO.
- 8K EPROM (8-bit) containing boot memory is connected. Since the program memory is 24-bit (3-byte) wide and since every 4th byte is not used, the boot

memory length is 2K.

- 8K external data memory (16-bit) is selected when address line A13 is zero. Hence the memory chip will be selected at any value below $2^{13} = 0x2000 = 8192$ beginning at absolute address 0.
- A/D converter is selected when both A2 and A13 are high. Hence the absolute address for A/D converter is $2^{13} + 2^2 = 0x2004 = 8196$.
- D/A converter is selected when both A1 and A13 are high. Hence the absolute address for D/A converter is $2^{13} + 2^1 = 0x2002 = 8194$.
- A parallel I/O port is found at 0x2001. This is selected when both A0 and A13 are High.
- This example system does not have any external program memory.

The System Specification File, EXMPL_2.SYS is shown below.

```
.SYSTEM                         EXAMPLE_2;        {System Name}
.ADSP2101;                                        {ADSP-2101 System}
.MMAP0;                                           {Enable Boot Loading}
.SEG/BOOT=0/ROM                 boot_mem[2048];   {Boot Segment}
.SEG/RAM/ABS=0x3800/DM/DATA     int_dm[1024];     {Internal Data Memory}
.SEG/RAM/ABS=0/DM/DATA          ext_dm[8192];     {External Data Memory}
.SEG/RAM/ABS=0/PM/DATA/CODE     int_pm[2048];     {Internal Program Memory}
.PORT/ABS=0x2004/DM             ad_converter;     {A/D Converter}
.PORT/ABS=0x2002/DM             da_converter;     {D/A Converter}
.PORT/ABS=0x2001/DM             io_port;          {Parallel I/O Port}
.ENDSYS;                                          {End of File}
```

The architecture description file, EXMPL_2.ACH, can now be generated by executing System Builder on EXMPL_2.SYS.

BLD21 EXMPL_2

A summary of the architecture created by the System Builder is shown below.

```
boot memory
0000-07ff [0800] Rom code/data BOOT_MEM
program memory
0000-07ff [0800] ram code/data 24 required bits in memory width INT_PM
data memory
0000-1fff [2000] ram data       EXT_DM
2001-2001 [0001] ram data       IO_PORT
2002-2002 [0001] ram data       DA_CONVERTER
2004-2004 [0001] ram data       AD_CONVERTER
3800-3bff [0400] ram data       INT_DM
```

Example 4-3:

In this example, a target system containing the ADSP-2101 is described as follows:

- Boot sequence disabled with MAP pin at HI level,
- 4K external data memory (ext_dmem) in data memory location 0x1000,
- 2K external program memory (ext_pmem) for both code and data beginning at 0x0000,
- Internal program memory is to be used for instruction code only,

- 2 A/D converters (adc_1 and adc_2) at data memory locations 0x3000 and 0x3001 respectively,
- One D/A converter (dac) at data memory location 0x3400.

The corresponding System Specification File, EXMPL_3.SYS is:

```
.SYSTEM                        EXAMPLE_3;      {System Name}
.ADSP2101;                                     {ADSP-2101 System}
.MMAP1;                                        {Disable Boot Loading}
.SEG/RAM/ABS=0x3800/DM/DATA    int_dmem[1024]; {Internal Data Memory}
.SEG/RAM/ABS=0x1000/DM/DATA    ext_dmem[4096]; {External Data Memory}
.SEG/RAM/ABS=0x3800/PM/CODE    int_pmem[2048]; {Internal Program Memory}
.SEG/RAM/ABS=0/PM/CODE/DATA    ext_pmem[2048]; {External Program Memory}
.PORT/DM/ABS=0x3000            adc_1;          {A/D Converter 1}
.PORT/DM/ABS=0x3001            adc_2;          {A/D Converter 2}
.PORT/DM/ABS=0x3400            dac             {D/A Converter}
.ENDSYS;                                       {End of File}
```

A summary of the architecture created by the System Builder is shown below.

```
boot memory
program memory
0000-07ff [0800] ram code/data 24 required bits in memory width EXT_PMEM
3800-3fff [0800] ram code       INT_PMEM
data memory
1000-1fff [1000] ram data       EXT_DMEM
3000-3000 [0001] ram data       ADC_1
3001-3001 [0001] ram data       ADC_2
3400-3400 [0001] ram data       DAC
3800-3bff [0400] ram data       INT_DMEM
```

4.2.3 System Architecture Exercises

The following exercises are designed to emphasize the description and use of the target system architecture. In each individual application, memory segments and I/O ports must be carefully described so that easier and better assembly language routines can be written.

1. Draw the program memory map, data memory map and write a System Specification file (.SYS) to describe the following ADSP-2101 hardware architecture. Be sure to include the on-chip memory.

   - Boot sequence enable,
   - 2K words of boot memory (boot_mem),
   - 4 A/D converters (ad_1, ad_2, ad_3 and ad_4) at data memory locations 0x2000, 0x2001, 0x2002 and 0x2003 respectively,
   - 4 D/A converters (da_1, da_2, da_3 and da_4) at data memory locations 0x1000, 0x1001, 0x1002 and 0x1003 respectively,
   - 2K external data memory (dm_ext) in data memory location 0x3000,
   - 1 I/O port (in_out) found in data memory location 0x3400.

2. A target system containing the ADSP-2101 is shown in Figure 4-3. Draw the program memory map, data memory map and write a System Specification file (.SYS) for this system. (Note: Data, Address, and Control signals are not shown.)

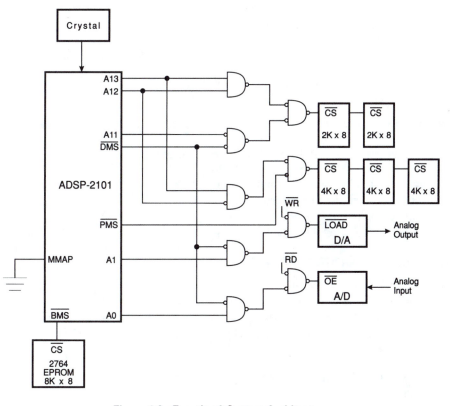

Figure 4-3: Exercise 2 System Architecture

4.3 SIMULATOR

The Simulator provides instruction-level simulation with a reconfigurable user interface. It is briefly described in Chapter 3 and a detailed description is available in the Cross Software Manual [2]. The Simulator is very useful not only in verifying the proper operation or debugging the improper one, but it is also useful in learning the architecture and instruction set of the ADSP-2101 chip. In this section, we will emphasize navigational and configurational aspects of the Simulator so that a user can get familiar with its environment. This is particularly useful when the ADSP-2101 instructions are studied in later subsections. More intricate functions of the Simulator will also be described briefly as they are needed in later chapters. We close this section with a few exercises on instruction sets using the Simulator environment.

The best way to learn the Simulator is to begin using it, experiment with different window operations and commands, and then build up custom screens with required fields. At some point, perhaps as soon as a few hours after the first encounter, the user can organize all the custom screens and commands into a clean set of external files. Thereafter, the user can invoke the Simulator with the appropriate startup file identified and the Simulator appears automatically in the desired configuration. After describing the necessary keywords for window configuration operations, we will provide a sample session of navigating through the Simulator. Although it is a very powerful tool in the entire development process, all its features cannot be learned unless we are debugging a complete program. Therefore we will emphasize some essential features of window commands and practice them through the study of ADSP-2101's instruction set.

The DOS command to invoke the Simulator along with its switches is described in Chapter 3. It is also given below:

```
sim2101 [-a arch_file] [-w window_file] [-s script_file]
```

The Simulator configures itself according to the target architecture file *arch_file* (default is 210x.ACH). It also uses two types of external configuration files: windows and scripts. Window files (default is DD.WIN) store look and layout of a particular set of windows. Scripts (default is STARTUP) are text files of command window inputs typically storing command aliases. In order to use the Simulator for program running or debugging, a Linker created symbol file (.SYM) must be available. It is created by using -g linker switch (described in Chapter 3).

4.3.1 Window Configurations

The Simulator is a window oriented interactive software tool. Up to ten windows can be opened at the same time. At any one time, there in always an active (highlighted) window. The first window , window #0, is called the Command Window. It is always present, cannot be deleted and it is used for accepting user input. Each newly opened window is assigned the next available window number.

The Window Configuration Menu is invoked by pressing Ctrl and W (depicted by ^W) keys simultaneously. It contains commands to open up a new window as well as moving, sizing, closing and hiding the active window. In the following discussion, note that we will use ^ to denote the Control key Ctrl and Enter↵ to denote the Enter key. Each command can be invoked by highlighting the corresponding command and pressing Enter↵ or by keying the first character of the command.

Window Operations

Open Windows

To open any window:
- Key ^W to display the Window Commands Menu.
- Select OPEN. A submenu of window selection will appear.
- Select the window you wish to open.
- The selected window will become the active window and will be displayed at the upper left corner of the screen

Select Active Window

Any newly opened window will become the active window automatically. There are two ways to select one of the opened windows as the active window.
1. Cycle through and set window active:

- Key ^Z to set the next window active in the numbered sequence.
2. Set window active by number:
   - Key ^X, followed by the window number and (Enter↵), to directly set the specified window active,
     e.g., ^X3 (Enter↵) will set window # 3 active.

Move Windows

To move any window:
- Select the window you want to move as the active window.
- Key ^W to display the Window Commands Menu.
- Select MOVE.
- Use arrow key to move the window and press (Enter↵) when done.
- An optional number, specifying the step size of the next arrow key, can be entered to speed up the moving process,
  e.g., 3 (←) will move the window to the left by 3 columns.

Size Window

To size any window:
- Select the window you want to size as the active window.
- key ^W to display the Window Commands Menu.
- Select SIZE.
- Use arrow key to size the window and press (Enter↵) when done.
- An optional number, specifying the step size of the next arrow key, can be entered to speed up the sizing process,
  e.g., 3 (←) will extend the window to left by 3 columns.

Close Window

To close any window:
- Select the window you want to close as the active window.
- Key ^W to display the Window Commands Menu.
- Select CLOSE.
- The active window will be closed.

Hide Window

To hide any window:
- Select the window you want to hide as the active window.
- Key ^W to display the Window Commands Menu.
- Select HIDE.
- The active window will be pushed to the bottom of the stack.

Decimal and Hex Display

To toggle between decimal and hex display
- Select the window you want to toggle the display.
- Key ^E to toggle between decimal and hex display.

Delete Window Field

To delete a field in the active window:
- Select the field by moving the cursor onto it.
- Key ^D.
- The field will disappear from the display.

Undelete Window Field

To undelete a field in the active window:
- Move the cursor to a blank location in the window.
- Key ^U.
- A menu of deleted fields for that window will appear.
- Select the desired field for undeletion.
- The deleted field will appear in that black location.

Move Window Field

To move a field in the active window:
- Select the field by moving the cursor onto it.
- Key ^Y to toggle on this function.
- Move the field, using the arrow keys, until it reaches the desired location.
- Key ^Y again or press Enter↵ to toggle off this function.

Table 4.1 below lists the keys and control key sequences which allows user to navigate between different windows or within the active window. These keystrokes can be used in *any window*.

Keystrokes	Description
^W	display main menu of window confuguration actions
Esc	exit a menu without making a selection
^Z	move to next (consecutively numbered) window
^X# Enter↵	move to window #
Arrow keys, PgUp, PgDn	Scroll through text in a window

Table 4-1: Window Navigation Controls ────────────────

4.3.2 A Sample Session

The tools for configuring an individual window are briefly described in the previous section. We now present a sample session to spell them out in greater detail with an example. In the process, we will use some window commands which are described in the next section. Invoke the Simulator by typing (we are using architecture description file generated in the example 4-1, however this is not essential and any .ACH can be used):

```
SIM21 -a exmpl.ach
```

A screen containing Command Window now appears.

Opening Windows

Key ^W to display the menu shown in Figure 4-4. Select OPEN by typing the letter "O" or pressing Enter↵ since OPEN is the default selection on this menu.

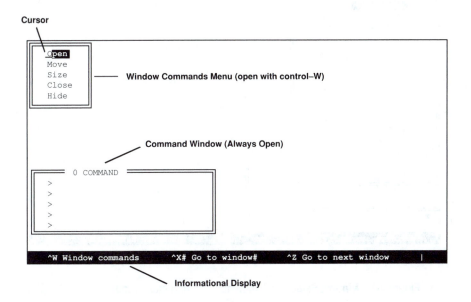

Figure 4-4: Main Menu For Configuring Windows

Doing this displays the window selection submenu shown in Figure 4-5. This session uses the register window. Select the register window by moving the cursor down with the array keys and pressing Enter↵ or by keying the menu letter "D".

Figure 4-6 shows the default register window layout. This is the starting point for rearranging the fields of this window.

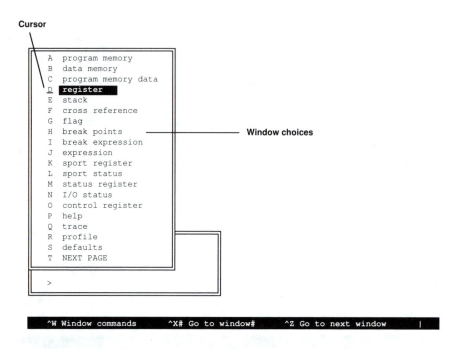

Cursor

```
A  program memory
B  data memory
C  program memory data
D  register
E  stack
F  cross reference
G  flag
H  break points                                    ────── Window choices
I  break expression
J  expression
K  sport register
L  sport status
M  status register
N  I/O status
O  control register
P  help
Q  trace
R  profile
S  defaults
T  NEXT PAGE

>
```

```
^W Window commands       ^X# Go to window#       ^Z Go to next window       |
```

Figure 4-5: Window Selection Submenu (with Register window selected)

```
1   REG    (REG_PRI,HEX)

ax0  uuuu    ar   uuuu    i0  uuuu    m0   uuuu    l0  uuuu    astat    00
ax1  uuuu    af   uuuu    i1  uuuu    m1   uuuu    l1  uuuu    mstat    00
ay0  uuuu                 i2  uuuu    m2   uuuu    l2  uuuu    sstat    55
ay1  uuuu                 i3  uuuu    m3   uuuu    l3  uuuu


mx0  uuuu    mr0  uuuu    i4  uuuu    m4   uuuu    l4  uuuu    ireq    000
mx1  uuuu    mr1  uuuu    i5  uuuu    m5   uuuu    l5  uuuu    imask    00
my0  uuuu    mr2  uu      i6  uuuu    m6   uuuu    l6  uuuu    icntl    uu
my1  uuuu    mf   uuuu    i7  uuuu    m7   uuuu    l7  uuuu


si   uuuu    sr0  uuuu    pc  0000    cntr uuuu              px   uu
se   uu      sr1  uuuu
sb   uu


cycle 00000000      irq2 00000000         dm_addr 0000      pm_addr 0000
```

Figure 4-6: Default Register Window Layout

Selecting , Deleting & Rearranging Fields In A Window

Note that only the windows displaying individual fields, like the register window, can be rearranged. The internal layout of memory windows and informational windows (like the breakpoint windows) cannot be altered.

For the purpose of illustration, let us assume that there are some registers in the processor which are never used. For example, only the SI register in the shifter is used (as a temparary holding register for signal data) and none of ALU registers are used. Likewise, only some of the DAG registers are used. Therefore we are going to delete unused registers from our display and rearrange the remaining registers for compactness.

The first step is to make the register window the active window, if it is not already. Key ^Z until it becomes the active window. The cursor now appears in the active window. Move the cursor with the arrow keys until it is over the SE field, one of the fields to be deleted. Key ^D to delete this field and it disappears from the display. Now move the cursor and delete the SB, SR0, and SR1 registers in the same way. Continue on to delete all of the ALU registers and the unneeded DAG registers: I2, I3, I5-7, M1-3 and M5-7, and L1-3 and L5-7. After these deletions, the register window looks like Figure 4-7.

```
┌══ 1   REG    (REG_PRI,HEX) ══════════════════════════════════════════┐
│                                                                      │
│                          i0   uuuu   m0   uuuu   l0   uuuu   astat  00│
│                          i1   uuuu                          mstat  00│
│                                                             sstat  55│
│                                                                      │
│                                                                      │
│    mx0  uuuu   mr0  uuuu  i4   uuuu   m4   uuuu   l4   uuuu   ireq   000│
│    mx1  uuuu   mr1  uuuu                                     imask  00│
│    my0  uuuu   mr2  uu                                       icntl  55│
│    my1  uuuu   mf   uuuu                                              │
│                                                                      │
│                                                                      │
│    si   uuuu              pc   0000   cntr uuuu              px uu    │
│                                                                      │
│                                                                      │
│    cycle 00000000     irq2 00000000     dm_addr 0000     pm_addr 0000 │
│                                                                      │
└──────────────────────────────────────────────────────────────────────┘
```

Figure 4-7: Example Register Window with some registers deleted

If a register is deleted accidently, it can easily be restored or "undeleted". Move the cursor to a blank spot in the window and key ^U. A menu drops down that lists all the deleted registers. Move the cursor along the menu and press (Enter↵) to restore any register. Press (Esc) to abort the operation. Note that the contents of the deleted register are also displayed. If a deleted register has a value other than undefined, the value is visible in this menu.

Now we can rearrange our pruned-down set of registers for a more compact display. Move the cursor to the MX0 register field. To move any field, select it for moving with ^Y, move it using arrow keys, and deselect it with another ^Y or [Enter↵]. Move the MX0 field up to the top line of the window this way. Repeating this way, we rearrange all the fields of this register window example until the entire window appears as shown in Figure 4-8.

```
┌─ 1   REG   (REG_PRI,HEX) ════════════════════════════════════════════════

  mx0   uuuu      mr0  uuuu     i0   uuuu    m0   uuuu    l0   uuuu    astat  00
  mx1   uuuu      mr1  uuuu     i1   uuuu                               mstat  00
  my0   uuuu      mr2  uu       i4   uuuu    m4   uuuu    l4   uuuu    sstat  55
  my1   uuuu      mf   uuuu


  si    uuuu                    ireq  000    imask  00    icntl  uu    px  uu


  cycle 00000000               pc   0000    cntr   uuuu        pm_addr  0000
  irq2  00000000                                               dm_addr  0000



```

Figure 4-8: Example Register Window with registers rearranged

Now we will resize the window outline, bringing up the bottom edge to make that space on the screen available for other windows. Key ^W to display the menu again and select size from it. As the up-arrow is pressed, the lower edge of the window moves up on the screen. Press [Enter↵] to end the sizing operation. The final version of this window should look like Figure 4-9.

Saving A Rearranged Screen

If we change the display without saving our custom register window, all the work creating it is lost. To save a new screen configuration with a desired set of windows opened, resized, and internally reconfigured, we must store the screen in a file. The Simulator command Y stands for the display and the greater than (>) and the less than (<) symbols are directional pipes. The display files are given default file extension .WIN. To save the current display (such as our example in Figure 4-9) enter this command in the command window:

y > 'myscreen'

This stores the current display configuration in the file MYSCREEN.WIN in the default directory. We can recall this or any other display configuration with the command

y < 'filename'

where *filename* is the main filename of a screen/window file.

```
┌─ 1   REG    (REG_PRI,HEX) ═══════════════════════════════════════════
│
│  mx0  uuuu    mr0  uuuu    i0   uuuu    m0   uuuu    l0   uuuu    astat  00
│  mx1  uuuu    mr1  uuuu    i1   uuuu                             mstat  00
│  my0  uuuu    mr2  uu      i4   uuuu    m4   uuuu    l4   uuuu    sstat  55
│  my1  uuuu    mf   uuuu
│
│
│  si   uuuu                 ireq  000    imask  00    icntl  uu    px  uu
│
│
│  cycle 00000000            pc  0000     cntr   uuuu      pm_addr  0000
│  irq2  00000000                                         dm_addr  0000
└──────────────────────────────────────────────────────────────────────
```

Figure 4-9: Final Register Window Arrangement

Closing Windows

During the course of a simulation, we may not have use of a particular window but would like to observe another one. This is done by closing one window and opening another one. In our session, we want to close our custom Register Window and keep only the Command Window (which cannot be closed). The first step is to make the register window the active window, if it is not already. Key ^Z until it becomes the active window. The cursor now appears in the active window. Key ^W to display the menu shown in Figure 4-4. Select CLOSE by typing the letter "C". The window now disappears from the screen and the cursor returns to the Command Window. Close any other opened windows.

Command Line Aliases

A command alias is a character string (plus any required arguments) which replaces one of the Simulators's native commands. The J command creates the alias. The syntax of this command is

j alias '*command*'

where *alias* is a symbol which will subsequently stand for the Simulator command enclosed in a single quotation marks.

For example, we can create a command that can call up file MYSCREEN.WIN, our previously saved display configuration. Remember that we used the command

y < '*myscreen*'

This may be inconvenient to type each time we want this display, especially because the "<" symbol requires the (Shift) key on most keyboards. We can alias this command with the new command string "VIEW" by entering the following:

j view '*y < "myscreen"*'

Note that the filename, *myscreen*, is enclosed with double quotes; nested quotation marks must be double, inside single. Some commands have arguments; these are passed using a dollar sign token as in $1,$2, etc.

4.3.3 Window Commands

In the sample session of the last section, we used a window command to save the created screen. These commands are entered in the Command Window. They can be categorized into different Simulator functions as discussed in Chapter 3. These functions include set-up functions, register functions, memory functions, control and debugging functions. In this section, we briefly describe commands from these functions.

Simulator Set-up Functions

These include loading the program to be simulated, opening I/O ports and associating data files with I/O ports and SPORTs for the purposes of simulating input and output data streams. Some miscellaneous set-up functions are also included.

Loading a Program

To load an ADSP-2101 .EXE and .SYM file, type

```
L 'filename'
```

where filename is the filename of the .EXE file. If .SYM is not found, a warning message appears and the Simulator will run without any labels and variables information. In other words, only the address will be displayed.

Opening and Closing an I/O Port

To open or close an I/O port type

```
O address [>'outfile.ext'] [< 'infile.ext']
```

Giving the O command with no optional parameters will close the port at the given address.

Opening and Closing a Serial Port

To open/close a serial port type

```
P 0 [>'outfile.ext'] [< 'infile.ext']
P 1 [>'outfile.ext'] [< 'infile.ext']
```

Giving the P command witn no optional parameters will close the corresponding serial port in either case. A 0 on the command line refers to SPORT0 while a 1 signifies the operation is done on SPORT1.

Other Defaults

There are a number of miscellaneous defaults for the operation of the Simulator. These can be changed in the defaults window. To change contents of the Default Window:
- Key ^W to display the Window Commands Menu.
- Select OPEN.
- Select Default Window.
- Modify the contents of the Default Window.

Simulator Inspecting and Altering Register Functions

The ADSP-2101 Simulator provides capabilities to inspect and modify different types of registers stored in the register, SPORT, status register, control register and stack windows. The syntax of inspecting a register, altering a register and undefining a register is described below.

Inspecting a Register

Use ? to inspect a register. For example,

```
? AX0
```

is used to inspect the content of the AX0 register

Altering a Register

You may use one of the two forms to alter the contents of any register:

1. Using the 'R' command

```
R Register Expression
```

where Register is the name of a processor register and expression is the value to be loaded into the register.

2. Direct assignment, e.g.,

```
AX0 = 0x002c
```

will assign the value 0x002c to register AX0

Undefining a Register

The command 'U' is used to undefine any given register, e.g.,

```
u AX0
```

will undefine register AX0

Simulator Inspecting and Altering Memory Functions

This section describes methods for viewing and altering specific locations in any of the ADSP-2101's program, data and boot memory spaces.

Inspecting a Memory Location

Use one of the following to inspect memory spaces.

1. (PgUp) and (PgDn) key
   - Open the desired memory window or make it active if already opened.
   - Use (PgUp) or (PgDn) key to scroll up and down inside the window
2. Use ^G key
   - Open the desired memory window or make it active if already opened.
   - Use ^G key and type in an address when prompted.
   - The window will display the input address.

3. The 'K' command in Command Window
   - Set the active window be the Command Window.
   - Execute the 'K' command, e.g.,

K 1 0x2000

will display the memory location 0x2000 of window # 1.
4. The 'D' command Window
   - Set the active window be the Command Window.
   - Execute the 'D' command, e.g.,

D *address* [> *'filename'*]
D *range*   [> *'filename'*]

where address or range specifies a location or range in memory. An optional redirection argument can be used to redirect the output to a disk file.

Tracking

Tracking is used to view the code and data memory while single stepping or running a program. The instruction being executed will be indicated when tracking is on in the Program Memory Window. Any change in data memory will be reflected immediately when tracking is turned on in the Data Memory Window. There are two ways to toggle the tracking feature:
1. The 'T' command in Command Window
   - Select the Command Window as the current window.
   - At the command prompt type

T *windownumber*

where *windownumber* is the number of the window you want to toggle the tracking effect.
2. The ^T key
   - Select the desired memory window as the current window.
   - Key ^T will toggle tracking effect in the current window.

Locating Symbols and Values

There are two ways to locate symbols and values in the program memory spaces.
1. The 'X' command
   - The 'X' command uses a symbol as argument and returns the address of the symbol if found. Note that symbols are case-sensitive, e.g.,

X *restart*

will return the address of *restart* if found.
2. The 'F' command
   - The 'F' command requires an address or range specifier as the first argument and an expression to be found as a second argument. It returns the address of the expression if found within the range, e.g.,

F pm[0]/100 *jump starter*

will search through the first 100 program memory spaces starting with the beginning of program memory location 0.

Altering a Memory Location

The 'E' command can be used for altering memory spaces. The syntax is as follows

E *address expression*

or

E *address [< 'filename']*

where *address* is a valid address or range specifier, *expression* is the value to be deposited in the prescribed range, and *filename* is a data file.

Altering Instructions

The 'A' command can be used to alter instructions in program memory spaces without converting the instruction into opcode (unlike the 'E' command), e.g.,

A pm[2] nop

will put a NOP instruction in pm[2] program memory location

Undefining Memory Location

The 'U' command can be used for undefining memory spaces. The syntax is as follows:

U *address*

where *address* is an individual address or an address range.

Control and Debugging Functions

Control functions include starting and stopping the execution of the program and resetting the simulated processor. Debugging functions include setting breakpoints, break conditions and watch points.

Control Functions

1. Chip reset and Reset
   - The command CR simulates a hardware reset which includes a boot loading sequence.
   - The command RE is the same as CR except it ignores the boot-loading sequence.
2. Single-step instructions
   The command

S *[stepsize]*

   executes the next *stepsize* instructions. It will go to the next instruction if the stepsize is omitted.
3. Running and Halting the processor
   - Use the 'G' command to run a program. It will stop when any key is pressed.

- The 'RUNFAST' command is the same as a 'G' command, except it does not stop when a key is pressed. The program will stop upon detection of an error or when a break point is encountered.

Break Functions

These commands halt execution. The Simulator break functions include break, break expressions, break changes and break ranges in program memory spaces

1. Setting break points and break ranges
    The command

B *address*

   or

B *range*

   halts at valid *address* or *range* specifier.
2. Viewing preset breakpoints
    The command

B *address*

   displays a listing of the breakpoints currently activated.
3. Break expressions
    The command

BE *expression*

   breaks whenever the *expression* is true.
4. Break on changes
    The command

BC *expression*

   breaks whenever the *expression* is changed.
5. Delete breakpoints
    The command

BD *address*

   deletes breakpoint at the given *address*.

Watch Functions

Watch functions display a message whenever the watchpoint criteria is met. These do not halt the execution of the program.

1. Setting watchpoints
    The command

W *address*

displays a message where *address* is a valid address or range specifiers.

2. Setting watch expressions

The command

WE *expression*

displays a message when *expression* is true.

3. Listing watchpoints and watch expressions

The command

W

displays a listing of the watchpoints and watch expresions.

4. Deleting watchpoints and watch expressions

The command

WD *watchnumber*

deletes a watch point or watch expression with the corresponding *watchnumber*.

Table 4-2 defines the control key sequences used only in particular windows. Table 4-3 lists a brief summary some useful commands and their arguments. A complete list of these commands is available in the *Cross Software Manual* [2].

Keystrokes	Description
^B	set breakpoint at cursor location in program memory window
^R	reset (delete) breakpoint at cursor location in program memory window
^D	delete field in active window
^U	undelete (restore) field in active window
^Y	move field in active window
^E	toggle numeric display of active window contents between HEX and DEC
^G	go to (prompted for address to be displayed)
^S	choose how many instructions to trace in trace window
^T	toggle tracking on/off in active window
^Q	quit simulator

Table 4-2: Window-Specific Control Key Sequences —————————————

4.3.4 Instruction Set Work-Out Using The Simulator

In Chapter 2, we described the complete instruction set of the ADSP-2101. To obtain a better understanding of each instruction, it is necessary to peek inside the processor and observe how the instruction works and how several registers are affected. This is possible using the Simulator because it allows us to execute individual instructions and then open different windows and look inside every field. One can also use the Emulator for this purpose but it is not as convenient as the Simulator. In this section, we will study some representative instructions from each group with emphasis on their effects on contents of various hardware elements. This will help us to understand the capabilities and limitations of those instructions. Some instructions like those in the Program Flow Control group are not possible to study using this approach because they require branching to another address location.

Keystrokes	Description
A *addr instr*	assemble instruction at address
B	list breakpoints, break expressions, and break change expressions
BD *addr* or *number*	delete breakpoint, break range, break change or break expression
CR	chip reset with boot page 0 load
G [*addr*]	start program execution
I *int# min max*	Cause periodic interrupt
J *symbol 'command'*	alias a command string
J	list alias
JD *symbol*	deelete alias
L *'file'*	load program into memory
O *addr* [<*'file'*] [>*'file'*]	open I/O port at addr and assign I/O file
O *addr*	close I/O port at addr
P SPORT#[<*'file'*][>*'file'*]	open serial port# 0 or 1 and assign I/O file
Q	quit simulator, with verification from user
R *regf expr*	set register equal to value of expression
RUNFAST	start program execution, no halt on key hit
S [*number*]	single step program execution
V *instr*	assemble and execute instruction
W *addr* or *range*	set watch point
WD *number*	delete watch point#
Y [>*'file'*][<*'file'*]	save/restore display configuration
Z [>*'file'*][<*'file'*]	save/restore Simulator state
?*expr*	evaluate expression

Table 4-3: Command Window Commands (Brief List) ——————————

Arithmetic on the ADSP-2101

To better understand the detailed discussion in the work-out, the user should first understand how the ADSP-2101 handles binary arithmetic. The ADSP-2101 is 16-bit, fixed-point machine. Special features support multiword arithmetic and block floating point. Most operations assume a twos-complement number while others assume an unsigned number or a simple binary string. In this section, we discuss the arithmetic used by each computational unit or operation.

Binary String

This is the simplest binary notation; sixteen bits are treated as a bit pattern. Examples of computation using this format are the logical operations: NOT, AND, OR and XOR. These ALU operations treat their operands as binary strings with no provision for sign bit or binary point placement.

Unsigned Numbers

Unsigned binary numbers may be thought of as positive, having nearly twice the magnitude of a signed number of the same length. The least significant words of multiple precision numbers are treated as unsigned numbers.

Signed Numbers: Twos-Complement

In discussions of ADSP-2101 arithmetic, "signed" refers to twos-complement. Most ADSP-2101 operations presume or support twos-complement arithmetic. The ADSP-2101 does not use signed-magnitude, ones-complement, BCD or excess-n formats.

Fractional Representation: 1.15

The ADSP-2101 is optimized for arithmetic values in a fractional binary format denoted by 1.15 ("one dot fifteen"). (Referred to in some contexts as 16.15 or Q15.) This is a fixed-point format. Used with the most significant bit (MSB) as a sign bit, the 1.15 means one sign bit and fifteen fractional bits representing values from -1 up to one least significant bit (LSB) less than +1.

ALU Arithmetic

All operations on the ALU treat operands and results as simple 16-bit binary strings, except the signed division primitive (DIVS). Various status bits treat the results as signed: the overflow (AV) condition code, and the negative (AN) flag.

The logic of the overflow bit (AV) is based on twos-complement. It is set if the MSB changes in a manner not predicted by the signs of the operands and the nature of the operation. For example, adding two positive numbers must generate a positive result; a change in sign bit signifies an overflow and sets AV. Adding a negative and a positive may result in either a negative or positive result, but cannot overflow.

The logic of the carry bit (AC) is based on unsigned-magnitude. It is set if a carry is generated from bit 16 (the MSB). The AC bit is most useful for the lower word portions of a multiword operation.

MAC Arithmetic

The multiplier produces results that are binary strings. The inputs are interpreted according to the information given in the instruction itself (signed times signed, unsigned times unsigned, a mixture or round). The 32-bit result from the multiplier is assumed to be signed, in that it is sign-extended across the full 40-bit width of the MR register set.

The ADSP-2101 supports two modes of format adjustment: the fractional mode for fractional operands, 1.15 format (1 signed bit, 15 fractional bits), and the integer mode for integer operands, 16.0 format. When multiplying 1.15 operands, the result is 2.30 (30 fractional bits). To correct this, in the fractional mode, a left shift occurs between the multiplier product (P) and the multiplier result register (MR). This shift (1 bit to the left) causes the multiplier result to be 1.31 which can be rounded to 1.15. In the integer mode, the left shift does not occur. For example, if the operands are in the 16.0 format, the 32-bit multiplier result would be in 32.0 format. A left shift would change the numerical representation resulting in an incorrect value.

Shifter Arithmetic

Most operations in the Shifter are explicitly geared to signed (twos-complement) or unsigned values: Logical shifts assume unsigned-magnitude or binary string values and Arithmetic Shifts assume twos-complement. The exponent logic assumes twos-complement numbers. It supports block floating point, which is also based on twos-complement fractions.

We will now study instructions from various groups using the Simulator. The most useful simulator command that we will use is the "V" command which performs line assembly and execution of an instruction. Also we will need the Register window which provides a display of all general purpose registers. To follow the exercises in this section, make sure that the Simulator is invoked properly and that the Command window is the active window. (In the following, we will use the Command Window prompt '>' to indicate a command line.)

ALU Group

This group contains arithmetic and logical operations which are easy to understand. First, open a Register window and Flag window, and then execute the following instructions.

```
> V AX0 = 3;
> V AY0 = 5;
> V AR = AX0 + AY0;
```

The first instruction puts decimal 3 in AX0 and the second one puts decimal 5 in AY0. Observe the register window as these instructions are executed. The default mode for the register contents is hex. Use ^E to toggle between hex and decimal mode. The third instruction adds AX0 and AY0 and puts the result in the AR register. The AR register now changes to 8 and all flags are cleared to 0. Now execute the following instruction which performs subtraction.

```
> V AF = AX0 - AY0;
```

The register AF changes to 0xFFFE (hex) or -2 (decimal). AN is set to 1 since result is a negative value and hence ASTAT becomes 0x02.

Now let us try some instructions which produce arithmetic overflow.

```
> V AX1 = 0X7FFF;
> V AY1 = 0X7FFF;
> V AR = AX1 + AY1;
```

AX1 and AY1 are set to 32767. AR register after addition changes to 0xFFFE which is 65534 but is interpreted as -2 in 2'complement arithmetic. This signals overflow and triggers the AN flag. Therefore AV and AN are set to 1 which changes ASTAT to 0x06.

```
> V AX0 = 0x9000;
> V AY0 = 0XA000;
> V AR = AX0 + AY0;
```

AX0 is set to -28672 and AY0 is set to -24576. Since 0x9000 + 0xA000 = 0x13000, AR changes to 0x3000 or 12288 but overflow and carry flags are triggered. AV and AC are set to 1 but AN changes to 0 (why?). Therefore ASTAT becomes 0x0C.

```
> V AX0 = 0;
> V AY0 = 0;
> V AR = AX0 + AY0 + C;
```

The last instruction adds AX0 and AY0 along with carry bit C. Since it was set by the last add instruction, AR is equal to 1 and all flags are cleared. ASTAT is also cleared to 0.

Try executing ALU instructions given in Chapter 2 and exercise the different ALU flags.

MAC Group

This group implements multiply and multiply/add operations. The important thing to learn here is how this multiplication is performed. The M_Mode bit (bit 4 in MSTAT register) controls whether the multiplication is in the default fractional format or in interger format. With the Register and Flag windows open, execute the following MAC instructions.

```
> V MX0 = 0X4000;
> V MY0 = 0X4000;
> V MR = MX0 * MY0 (ss);
```

The MAC assumes data in a 1.15 fractional format. The first two instructions therefore put 0.5 in MX0 and MY0 registers. Now multiplication of two 1.15 format numbers is a number in 2.30 format. However, MAC changes this format by an automatic 1-bit left shift in the default mode and the result is a number in 1.31 format. The third instruction multiplies MX0 and MY0 (where both are assumed to be in signed form) and puts the result in MR. MR is a 40 bit register which is displayed as two 16-bit MR0 and MR1 registers and a 8-bit MR2 register. The result of the above operation is illustrated below. No flags are set in this operation.

```
    MX0        = 0.100 0000 0000 0000
*   MY0        = 0.100 0000 0000 0000
------------------------------------
    MX0 * MY0 = 0.001 0000 0000 0000 0000 0000 0000 0000
After 1-bit shift to the left
    MX0 * MY0 = 0.010 0000 0000 0000 0000 0000 0000 0000  (0.25 decimal)

=> MR2 = 0x00
=> MR1 = 0x2000
=> MR0 = 0x0000
```

Let us now try the following instructions.

```
> V MX1 = 2;
> V MY1 = 3;
> V MR = MX1 * MY1 (SS);
```

Note that decimals 2 and 3 are put in MX1 and MY1 respectively. The result however is not the expected integer 6 but 12 because of the default fractional format as explained below. Also note that the result is interpreted not as an integer 12 but some fractional number (which one?).

```
    MX1        = 0.000 0000 0000 0010
*   MY1        = 0.000 0000 0000 0011
------------------------------------
    MX1 * MY1 = 0.000 0000 0000 0000 0000 0000 0000 0110
After 1-bit shift to the left
    MX1 * MY1 = 0.000 0000 0000 0000 0000 0000 0000 1100

=> MR2 = 0x00
=> MR1 = 0x0000
=> MR0 = 0x000C
```

Now we will turn on the integer multiplication mode. This is done by enabling M_MODE bit. Execute the following instructions.

```
> V ENA M_MODE;
> V MR = MX1 * MY1 (SS);
```

M_MODE bit (bit 4 of the MSTAT register) is set after execution of the ENA (Enable) instruction. It forces subsequent MAC instructions to suppress automatic left shift of the product and implements integer product. Multiplication of MX1 and MY1 now results in 6.

```
    MX1        = 0.000 0000 0000 0010
*   MY1        = 0.000 0000 0000 0011
-----------------------------------
    MX1 * MY1 = 0.000 0000 0000 0000 0000 0000 0000 0110

=> MR2 = 0x00
=> MR1 = 0x0000
=> MR2 = 0x0006
```

Let us disable the integer mode and study some other aspects of MAC instructions. Try the following.

```
> V DIS M_MODE;
> V MR1 = 0X7FFF;
> V MR0 = 0XFFFF;
> V MR = MR + MX0 * MY0 (ss);
```

MR is set to the highest positive number in 1.31 format which is ≈ 1. Therefore any addition to it will cause an overflow. The last instruction is a multiply/add instruction which multiplies MX0 (0x4000) with MY0 (0x4000), adds the result to MR and stores the new value in MR.

```
    MX0        = 0.100 0000 0000 0000
*   MY0        = 0.100 0000 0000 0000
-----------------------------------
    MX0 * MY0 = 0.010 0000 0000 0000 0000 0000 0000 0000 (after 1-bit shift)
+   MR         = 0.111 1111 1111 1111 1111 1111 1111 1111
-----------------------------------------------------
    MR         = 1.001 1111 1111 1111 1111 1111 1111 1111

=> MR2 = 0x00
=> MR1 = 0x9FFF
=> MR0 = 0xFFFF
```

The multiply overflow bit MV is set to 1 and ASTAT changes to 0x40. This overflow bit can be used to take some action. Execute the following control flow command.

```
> V if MV SAT MR;
```

Since MV is set, MR will be saturated and set to the maximum possible fraction.

```
MR2 = 0x00
MR1 = 0x7FFF
MR2 = 0xFFFF
```

Try executing some more MAC instructions and learn about different formats and the saturation mode.

Shifter Group

This group performs arithmetic/logic shifts and operations pertaining to floting-point arithmetic. These instructions are fairly easy to understand. With the Register and Flag windows open, execute the following instructions.

```
> V SI = 0x1111;
> V SR = LSHIFT SI by 2 (HI);
```

The first instruction puts hex 1111 in register SI. The second instruction, LSHIFT, is a logical shift (not left shift) instruction in which 32-bit output field SR is zero-filled from right for left shift and zero-filled from left for right shift. The above instruction shifts SI logically to the left by 2 bits (multiply the number by 2^{+2}), hence SI = 0x4444. The result is stored in the high 16 bits of the SR register, i.e., SR1 = 0x4444 and SR0 = 0x0000

```
> V SR = LSHIFT SI by -2 (HI);
```

Now SI is shifted logically 2 bits to the right (multiply the number by 2^{-2}) and the result is stored in high registers of SR, hence SR1 = 0x0444 and SR0 = 0x4000

```
> V SR = SR or LSHIFT SI by 1 (LO);
```

When SI is logically shifted to left by 1 bit, the result is 0x2222. After the OR operation, SR becomes

```
        SR = 0x0444 4000
OR           0x0000 2222
     --------------------
        SR = 0x0444 6222

=> SR1 = 0x0444
=> SR0 = 0x6222
```

Next, we will consider an arithmetic shift in which a 32-bit output field SR is zero-filled from right for left shift and sign-extended to the left for right shift.

```
> V SI = 0xC000;
> V SR = ASHIFT SI by -8 (HI);
```

In the first instruction SI is set to 0xC000. The second instruction shifts SI arithmetically 8 bits to the right and the result is stored in high 16 bits of SR register. The result is 0xFFC0.

```
         SI = 1100 0000 0000 0000
ASHIFT -8     1111 1111 1100 0000

=> SR1 = 0xFFC0
=> SR0 = 0x0000
```

The Shifter also performs exponent extraction and normalization operations needed in floating-point arithmetic. Consider the following instructions.

```
> V SI = 0x0FFF;
> V SE = EXP SI (HI);
```

The first instruction sets SI to 0x0FFF. In 1.15 format, SI = 0.000 1111 1111 1111 . There are 3 zeroes to the right of the binary point. The second instruction extracts the exponent for normalization. Therefore, SE = -3 (decimal) or 0xFD. Note that SE is set to the negative of the exponent needed for normalization. In the following instruction

```
> V SR = NORM SI (HI);
```

since SE = -3, the content of SI are logically shifted to the left 3 bits and the result is stored in the high 16 bits of the SR register, hence
SR1 = 0x7FF8 and SR0 = 0x0000

Try executing some more shifter instructions given in Chapter 2.

DAG Group

This set of instructions perform data move operations by generating appropriate addresses. The important thing to learn here is how addresses are generated by I register and modified by M and L registers. In this section we will move values into data memory. Therefore open a Register as well as a Data Memory window and execute the following instructions.

```
> V I0 = 4;
> V M1 = 3;
> V L0 = 8;
> E DM[0]/8 0x1234;
```

The first three instructions move values into I0, M1 and L0 registers. The last command "E" enters 0x1234 into 8 data memory locations beginning at 0. Threfore the contents of dm[0] through dm[7] are equal to 0x1234. Now execute the following data memory read instruction.

```
> V AX0 = DM(I0, M1);
```

The index register I0 is equal to 4 before the execution, therefore DM[4] (= 0x1234) is read into AX0. After memory read, I0 = (I0 + M1) mod L0. Hence I0 becomes (3 + 4) mod 8 = 7.

```
> V AY0 = DM(I0, M1);
```

Now AY0 = DM[7] = 0x1234, while I0 changes to (7 + 3) mod 8 = 2. Threfore the next time data memory is read, I0 will point towards memory location 2. This is called circular buffering and it is an important operation in digital signal processing. It should be noted that the buffer length register L# must go with the index register I#, while the modify register M% can be anyone from the set.

Data Address Generator 1 also has a capability to provide bit-reverse logic intended for use in fast Fourier transform computations where inputs are supplied or outputs generated in bit-reversed order. The pivot point for the reversal is the midpoint of the 14-bit address, between bits 6 and 7. This is illustrated in the following chart.

Individual DMA lines (DMA_N)

Normal order	13	12	11	10	09	08	07	06	05	04	03	02	01	00
Bit-reversed	00	01	02	03	04	05	06	07	08	09	10	11	12	13

Bit-reversed addressing mode is enabled or disabled by setting bit 1 in the mode status register (MSTAT). Try the following instructions.

```
>V ENA BIT_REV;
>V SI = DM(I0,M1);
```

The first instruction enables bit reversal. When enabled, all addresses generated using index registers I0-3 are bit-reversed upon output. Since I0 = 2, the generated address is 4096 (why?). Hence DM[4096] is read into SI. However, after memory read, I0 changes to $(2 + 3)$ mod 8 = 5. In other words, the modified value stored back after post-update remains in normal order.

Try executing some more DAG instructions given in Chapter 2.

4.4 IN-CIRCUIT EMULATOR (EZ-ICE)

The "EZ-ICE" In-Circuit Emulator provides a testing and debugging environment to check digital signal processing application programs through a menu-driven, easy-to-use interface. Even though the Simulator can support debugging facilities, real-time operational testing can only be done on an Emulator. It is an essential tool in an industrial setting. In an educational (laboratory) environment in which the emphasis is on the basics of the digital signal processor programming, the Emulator is useful only as a final link in the development. This section therefore deals with some basic operations of the EZ-ICE system and describes the Command Menu structure. Detailed information can be obtained in the EZ-ICE manual [3].

The target system for the EZ-ICE (also referred to as the Probe) should also be available before programs can be tested. We assume that such a target system is the EZ-LAB system described in Chapter 1. To communicate with the EZ-ICE through the PC, a terminal emulator program is required. For the purpose of discussion we also assume that a program called PROCOMM is available. However this is not essential and any communication program capable of emulating a VT100 type terminal can be used.

4.4.1 Hardware Operations of EZ-ICE

Consult your lab supervisor for power supply and PC connections to EZ-ICE and EZ-LAB boards. General Power up and power down sequences are described below.

Power Up Sequences

- Start terminal emulation software (e.g. Procomm) on your PC.

- If separate power supplies are used for EZ-LAB and EZ-ICE, then
  -Turn on power supply for EZ-ICE,
  -Press [Enter↵] on your PC,
  -When advised to turn on power of the target system(e.g. EZ-LAB), do so.

- If a single power supply is used for both EZ-ICE and EZ-LAB, then
  -Turn on the power supply,
  -Press [Enter↵] on your PC,
  -Press another [Enter↵] to continue.

The above procedures are important in order to safeguard the EZ-LAB and EZ-ICE circuitry.

Power Down Sequences

- If separate power supplies are used for EZ-LAB and EZ-ICE, remove power from the EZ-LAB FIRST before removing power from EZ-ICE.

- If a single power supply is used, turn it off when you are done.

4.4.2 Invoking Emulator Firmware

The Emulator has its own software stored on a ROM which contains diagnostic tests and its menu system. The power up sequence resets the Emulator or pressing the RESET switch on the board produces the same result.

Initial Display

After EZ-ICE is reset, press the PC's Enter↵ key to obtain the following initial display.

```
"EZ-ICE" ADSP-2101 IN-CIRCUIT EMULATOR  -  Analog Devices, Inc.   Version
1.1

RS-232 COMMUNICATIONS ESATBLISHED AT 9600 BAUD.

TURN ON POWER FOR TARGET SYSTEM NOW.
-- Hit Carriage Return To Continue --

> SELF TEST IN PROGRESS!

  1) HOST RAM TEST
     HOST RAM OK!

  2) ADSP-2101 FUNCTION TEST
     ADSP-2101 OK!
 ** MMAP Configuration Is Set For Internal Program Memory At Location 0000
**

 Do You Wish To Use Overlay Memory (Y or N)?
     * * I M P O R T A N T * *
     Target Memory MUST Be Removed From Your Circuit If You Use Overlay Memory!
```

If the answer is 'N', the EZ-ICE menu will appear. If the answer is 'Y', the software will invoke overlay memory test.

Overlay Memory Test

If the answer to the initial display is 'Y', overlay memory test will be invoked. The overlay memory test screen should look as follows:

```
  3) OVERLAY MEMORY TEST
     * * I M P O R T A N T * *
     Target Memory MUST Be Removed From Your Circuit For This Test!
     This Test Should Only Be Run If You Intend To Use Overlay Memory,
     Since Overlay Memory Is Enabled And Left Enabled After The Test.
        (See Manual For Jumper Configuration)
     Type TEST To Proceed With Test, Hit Any Other Key To Omit Test!
     >
```

If you type anything other than "test", overlay memory test will be skipped. Otherwise, the following question will be displayed on the screen.

What Is The Position Of Jumper JP1   (1, 2 or 3)?

Type in the Position of Jumper JP1 and the overlay memory test begins. If the overlay memory test fails, verify that the jumper position you entered and the actual jumper position are the same. Also, make sure the target memory that the overlay memory replaces has been removed.

4.4.3 EZ-ICE Basic Command Menu

All emulator operations originate either from the keyboard input at this level, or from one of the subordinate displays or menus reached through it. The Basic Command Menu looks as follows:

```
=================================================
|      "E Z - I C E"    ADSP-2101   EMULATOR    |
|                 Basic Command Menu            |
|                                               |
|      Read/Write Registers                     |
|      Read Data Memory                         |
|      Write Data Memory                        |
|      Read Program Memory                      |
|      Write Program Memory                     |
|      Reset Processor                          |
|      Single Step                              |
|      Set, Clear, List Breakpoints             |
|      Run Until Breakpoint                     |
|      Boot From External EPROM                 |
|      Download From Host                       |
|      Upload To Host                           |
|      Display Stacks                           |
|                                               |
=================================================
```

You can control the menu selection by positioning the cursor on the item desired using the Up and Down Arrow keys and pressing ⏎Enter⏎.

Read/Write Registers

This is the first entry on the Basic Commands Menu. You can view and modify the contents of processor registers manually. The Read/Write Registers display is shown as follows:

```
=========================================================================
| "EZ-ICE"  ADSP-2101  Register Display  (Values Are Displayed in HEX)  |
|           ALU                                      Address Generator # 1 |
|AX0 0000                                   I0 0000   M0 0000   L0 0000|
|AX1 0000                AR 0000   AN 0  AC 0   I1 0000   M1 0000   L1 0000|
|AY0 0000                AF 0000   AV 0  AQ 0   I2 0000   M2 0000   L2 0000|
|AY1 0000                AZ 0  AS 0   I3 0000   M3 0000   L3 0000|
|         Multiplier-Accumulator                                          |
|MX0 0000                                      Address Generator # 2 |
|MX1 0000   MR2 0000  MR1 0000      MR0 0000    I4 0000   M4 0000   L4 0000|
|MY0 0000             MF   0000                 I5 3000   M5 0000   L5 0000|
|MY1 0000                          MV 0         I6 3000   M6 0000   L6 0000|
|         Shifter                               I7 07FF   M7 0000   L7 0000|
|SI  0000   SE 0000   SR1 0000      SR0 0000                              |
|           SB 0000            SS 0                PX 00CF                |
|                                                                         |
|CNTR 0000 Astat 0000  Mstat 0000  Sstat 0055  Imask 0000  Icntl 0017   |
|  PC 0000            RX0 006A     TX0 0055    RX1 006A    TX1 0055   |
|           Single Step                                                   |
|           Display Control Registers                                     |
|           Return To Basic Command Menu                                  |
=========================================================================
```

        ** Place Cursor Over First Digit Of Data Field For Data Entry **

The keyboard keys you may use in this display are:

Up Arrow	Move up one line
Down Arrow	Move down one line
Left Arrow	Move left one column
Right Arrow	Move right one column
Tab	Move 10 positions to right on a line
^A	To move to the first entry (Ax0)

There are three options associated with this display: Single Step, Display Control Register and Return to Basic Command Menu.

Single Step

This option allows execution of instructions while monitoring the resulting states of the registers. It is similar to the Simulators single stepping of instructions.

Display Control Registers

This option will display the contents of the 17 memory-mapped Control Registers found from 0x3FEF to 0x3FFF in data memory. We can view and modify the contents of all the control registers. This display is shown as follows:

```
===================================================================
| ADSP-2101  Control  Register Display (Values Displayed In HEX) |
|                                                                 |
|     System Control Register                          001F       |
|     Data Memory Wait State Control Register          7FFF       |
|     TPERIOD Period Register                          0000       |
|     TCOUNT  Counter Register                         0000       |
|     TSCALE  Scaling Register                         0000       |
|     SPORT0 Multichannel RCV Word Enable Reg. (MS)    80FD       |
|     SPORT0 Multichannel RCV Word Enable Reg. (LS)    0000       |
|     SPORT0 Multichannel TX Word Enable Reg. (MS)     0080       |
|     SPORT0 Multichannel TX Word Enable Reg. (LS)     FF7F       |
|     SPORT0 Control Register                          0000       |
|     SPORT0 SCLKDIV Serial Clock Divide Modulus       0000       |
|     SPORT0 RFSDIV Receive Frame Sync Divide Modulus  0000       |
|     SPORT0 Autobuffer Control Register               0000       |
|     SPORT1 Control Register                          0000       |
|     SPORT1 SCLKDIV Serial Clock Divide Modulus       0000       |
|     SPORT1 RFSDIV Receive Frame Sync Divide Modulus  0000       |
|     SPORT1 Autobuffer Control Register               0000       |
|                                                                 |
|            Return To Data Register Display                      |
===================================================================
** Place Cursor Over First Digit Of Data Field For Data Entry **
```

Return to Basic Command Menu

Pressing ⌷Enter↵⌷ on this selection will display Basic Command Menu.

Read Data Memory

This entry allows us to view the 16-bit contents of the data memory location plus ten locations above and below the address. For example, to read data memory at hex location 3810, select the Read Data Memory entry and enter the location as shown below.

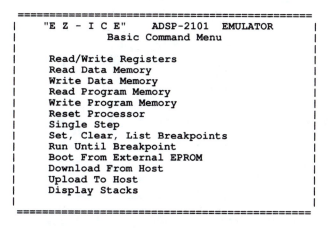

```
================================================
|      "E Z - I C E"    ADSP-2101   EMULATOR     |
|                  Basic Command Menu            |
|                                                |
|      Read/Write Registers                      |
|      Read Data Memory                          |
|      Write Data Memory                         |
|      Read Program Memory                       |
|      Write Program Memory                      |
|      Reset Processor                           |
|      Single Step                               |
|      Set, Clear, List Breakpoints              |
|      Run Until Breakpoint                      |
|      Boot From External EPROM                  |
|      Download From Host                        |
|      Upload To Host                            |
|      Display Stacks                            |
|                                                |
================================================

ENTER HEX MEMORY ADDRESS: 3810
```

After specifing the address 3810, the corresponding data memory locations are displayed as follows:

```
    HEX Addr.   DM Value (HEX)
        3806        400C
        3807        0000
        3808        0000
        3809        0100
        380A        0000
        380B        0804
        380C        8300
        380D        0000
        380E        0000
        380F        0000
   >    3810        8001
        3811        4080
        3812        0004
        3813        0000
        3814        0000
        3815        0000
        3816        1001
        3817        0000
        3818        0100
        3819        0004
        381A        0000

     ** Hit Carriage Return To Return To Basic Command Menu **
```

Using arrow keys, adjacent data memory locations can also viewed.

Write Data Memory

We can write a value to one or more data memory addresses by selecting the Write Data Memory entry in the Basic Command Menu. For example, to write hex 1234 in 16 data memory locations beginning at hex 3800, enter the values as shown below.

```
===================================================
|        "E Z - I C E"      ADSP-2101   EMULATOR        |
|                    Basic Command Menu                 |
|                                                       |
|        Read/Write Registers                           |
|        Read Data Memory                               |
|        Write Data Memory                              |
|        Read Program Memory                            |
|        Write Program Memory                           |
|        Reset Processor                                |
|        Single Step                                    |
|        Set, Clear, List Breakpoints                   |
|        Run Until Breakpoint                           |
|        Boot From External EPROM                       |
|        Download From Host                             |
|        Upload To Host                                 |
|        Display Stacks                                 |
|                                                       |
===================================================

    ENTER HEX MEMORY ADDRESS:      3800
    ENTER HEX DATA VALUE:          1234
    ENTER HEX NUMBER OF LOCATIONS: 10
```

Then, the modified data memory locations will be displayed as follows.

```
    HEX Addr.   DM Value (HEX)
        37F6        FFFF
        37F7        FFFF
        37F8        FFFF
        37F9        FFFF
        37FA        FFFF
        37FB        FFFF
        37FC        FFFF
        37FD        FFFF
        37FE        FFFF
        37FF        FFFF
    >   3800        1234
        3801        1234
        3802        1234
        3803        1234
        3804        1234
        3805        1234
        3806        1234
        3807        1234
        3808        1234
        3809        1234
        380A        1234
```

        ** Hit Carriage Return To Return To Basic Command Menu **

Read Program Memory

The Read Program Memory command on the Basic Command Menu allows us to view the 24-bit contents of the program memory location selected, plus ten locations above and below the address. To read program memory location at hex 20, select the Read Program Memory entry as shown below.

```
==================================================
|      "E Z - I C E"     ADSP-2101   EMULATOR      |
|                Basic Command Menu                |
|                                                  |
|      Read/Write Registers                        |
|      Read Data Memory                            |
|      Write Data Memory                           |
|      Read Program Memory                         |
|      Write Program Memory                        |
|      Reset Processor                             |
|      Single Step                                 |
|      Set, Clear, List Breakpoints                |
|      Run Until Breakpoint                        |
|      Boot From External EPROM                    |
|      Download From Host                          |
|      Upload To Host                              |
|      Display Stacks                              |
|                                                  |
==================================================

ENTER HEX MEMORY ADDRESS: 20
```

Then, the program memory address, plus ten location above and below it, the Program Counter (PC) and any pre-defined breakpoints (BK) will be displayed as follows.

```
      HEX Addr.   PM Value (HEX)    Disassembled Contents      MMAP = 0
        0016        0A001F          RTI
        0017        0A001F          RTI
        0018        0A001F          RTI
        0019        0A001F          RTI
        001A        0A001F          RTI
        001B        0A001F          RTI
        001C        1C097F          CALL  0x0097
 <PC>   001D        378000          I0=0x3800
        001E        340014          M0=0x0001
        001F        341008          L0=0x0100
   >    0020        379052          I2=0x3905
        0021        340006          M2=0x0000
        0022        34000A          L2=0x0000
        0023        381000          I4=0x0100
        0024        380014          M4=0x0001
        0025        381008          L4=0x0100
        0026        382001          I5=0x0200
        0027        381009          L5=0x0100
        0028        383002          I6=0x0300
        0029        38100A          L6=0x0100
        002A <BK>   384003          I7=0x0400

       ** Hit Carriage Return To Return To Basic Command Menu **
```

Write Program Memory

The Write Program Memory entry allows us to modify the contents of any existing RAM location in program memory with a six-digit hexadecimal value. To change the instruction of a program memory location select the Write Program Memory entry in the Basic Command Menu. The EZ-ICE firmware will request for the hexadecimal location of the program memory we want to modify. Then, a six-digit instruction encoding number is asked for input. Finally, we will be prompted for the number of program memory addresses we want filled with this value.

```
==================================================
|       "E Z - I C E"    ADSP-2101    EMULATOR       |
|                  Basic Command Menu               |
|                                                   |
|       Read/Write Registers                        |
|       Read Data Memory                            |
|       Write Data Memory                           |
|       Read Program Memory                         |
|       Write Program Memory                        |
|       Reset Processor                             |
|       Single Step                                 |
|       Set, Clear, List Breakpoints                |
|       Run Until Breakpoint                        |
|       Boot From External EPROM                    |
|       Download From Host                          |
|       Upload To Host                              |
|       Display Stacks                              |
|                                                   |
==================================================
```

```
ENTER HEX MEMORY ADDRESS:      A
ENTER HEX DATA VALUE:            1800C0
ENTER HEX NUMBER OF LOCATIONS:4
```

As soon as you press ⌈Enter↵⌋ on the last entry of the Write Program Memory, the modified program memory will be displayed as follows.

```
      HEX Addr.   PM Value (HEX)    Disassembled Contents    MMAP = 0
     <PC> 0000       1801CF         JUMP   0x001C
          0001       0A001F         RTI
          0002       0A001F         RTI
          0003       0A001F         RTI
          0004       1807AF         JUMP   0x007A
          0005       0A001F         RTI
          0006       0A001F         RTI
          0007       0A001F         RTI
          0008       0A001F         RTI
          0009       0A001F         RTI
        > 000A       1800C0         IF EQ JUMP   0x000C
          000B       1800C0         IF EQ JUMP   0x000C
          000C       1800C0         IF EQ JUMP   0x000C
          000D       1800C0         IF EQ JUMP   0x000C
          000E       0A001F         RTI
          000F       0A001F         RTI
          0010       0A001F         RTI
          0011       0A001F         RTI
          0012       0A001F         RTI
          0013       0A001F         RTI
          0014       0A001F         RTI

        ** Hit Carriage Return To Return To Basic Command Menu **
```

Reset Processor

The Emulator can be re-initialized upon selection of this entry. Upon completion of the reset, the "Processor Reset Complete!" message is displayed briefly.

```
===================================================
|     "E Z - I C E"    ADSP-2101   EMULATOR     |
|              Basic Command Menu               |
|                                               |
|      Read/Write Registers                     |
|      Read Data Memory                         |
|      Write Data Memory                        |
|      Read Program Memory                       |
|      Write Program Memory                     |
|      Reset Processor                          |
|      Single Step                              |
|      Set, Clear, List Breakpoints             |
|      Run Until Breakpoint                     |
|      Boot From External EPROM                 |
|      Download From Host                       |
|      Upload To Host                           |
|      Display Stacks                           |
|                                               |
===================================================

              Processor Reset Complete !
```

Single Step

We can perform a single step operation while viewing the registers in the Read/Write Registers display. We can also do this from the Basic Command Menu by selecting the Single Step entry:

```
===================================================
|     "E Z - I C E"    ADSP-2101   EMULATOR     |
|              Basic Command Menu               |
|                                               |
|      Read/Write Registers                     |
|      Read Data Memory                         |
|      Write Data Memory                        |
|      Read Program Memory                       |
|      Write Program Memory                     |
|      Reset Processor                          |
|      Single Step                              |
|      Set, Clear, List Breakpoints             |
|      Run Until Breakpoint                     |
|      Boot From External EPROM                 |
|      Download From Host                       |
|      Upload To Host                           |
|      Display Stacks                           |
|                                               |
===================================================

   Addr: 0000    Instr:  JUMP  0x001C
```

Set, Clear, List Breakpoints

EZ-ICE is capable of supporting as many as 16 addressed breakpoints in program memory. Breakpoints can be set, cleared and/or listed from the Breakpoint Display which is reached from the Basic Command Menu by selecting the Set, Clear, List Breakpoints entry.

The presence of a breakpoint is shown as "<BK>" in the Read Program Memory display. A snapshot of the Breakpoint Display is shown below.

```
Breakpoint Number        Breakpoint Address (HEX)

        0                        00FC
        1                        0180
        2                        0232
        3                         —
        4                         —          Enter Breakpoint Address
        5                         —             At Cursor Position
        6                         —
        7                         —
        8                         —
        9                         —
        A                         —
        B                         —
        C                         —
        D                         —
        E                         —
        F                         —


  ** To Delete A Breakpoint, Type X Followed By The Breakpoint Number **

    **   Hit Carriage Return To Return To Basic Command Menu    **
```

To add a breakpoint, type in a four-digit hexadecimal address followed by ⌑Enter⌑. To remove a breakpoint, type in "x#" where "#" is the number of the breakpoint you want to remove.

Run Until Breakpoint

The Run Until Breakpoint entry on Basic Command Menu causes the Emulator to execute instructions from the current program counter address until the next breakpoint is encountered. However, the execution of instructions can be stopped by pressing any key on the keyboard. A snapshot of the Processor Running Screen is shown below.

```
================================================
|    "E Z - I C E"    ADSP-2101   EMULATOR     |
|            Basic Command Menu                |
|                                              |
|      Read/Write Registers                    |
|      Read Data Memory                        |
|      Write Data Memory                       |
|      Read Program Memory                     |
|      Write Program Memory                    |
|      Reset Processor                         |
|      Single Step                             |
|      Set, Clear, List Breakpoints            |
|      Run Until Breakpoint                    |
|      Boot From External EPROM                |
|      Download From Host                      |
|      Upload To Host                          |
|      Display Stacks                          |
|                                              |
================================================

PROCESSOR RUNNING!
```

Boot From External EPROM

EZ-ICE has the capability of transferring any of up to 8 pages of external boot EPROM in the target system into the 2K, 24-bit words of program memory RAM within the Emulator. Booting is available only if the MMAP pin is pulled to low (i.e. MMAP = 0). An error message will result upon selection of this entry if MMAP = 1. A snapshot of the Booting Screen is shown below.

```
==================================================
|      "E Z - I C E"     ADSP-2101   EMULATOR    |
|                 Basic Command Menu             |
|                                                |
|       Read/Write Registers                     |
|       Read Data Memory                         |
|       Write Data Memory                        |
|       Read Program Memory                      |
|       Write Program Memory                     |
|       Reset Processor                          |
|       Single Step                              |
|       Set, Clear, List Breakpoints             |
|       Run Until Breakpoint                     |
|       Boot From External EPROM                 |
|       Download From Host                       |
|       Upload To Host                           |
|       Display Stacks                           |
|                                                |
==================================================

   Enter Boot Page Number : 5
   ** Boot From External EPROM Page 05 Complete **
    Hit Carriage Return To Continue
```

Download From Host

Selection of this entry enables downloading of Memory Image Files (.EXE) from the PC to EZ-ICE. The following is the Download From Host screen.

```
==================================================
|      "E Z - I C E"     ADSP-2101   EMULATOR    |
|                 Basic Command Menu             |
|                                                |
|       Read/Write Registers                     |
|       Read Data Memory                         |
|       Write Data Memory                        |
|       Read Program Memory                      |
|       Write Program Memory                     |
|       Reset Processor                          |
|       Single Step                              |
|       Set, Clear, List Breakpoints             |
|       Run Until Breakpoint                     |
|       Boot From External EPROM                 |
|       Download From Host                       |
|       Upload To Host                           |
|       Display Stacks                           |
|                                                |
==================================================

         Start Download From Host Now!

   Type CNTL C To Abort
```

If MMAP = 0, any boot kernel within location 0x0000 and 0x07FF will be loaded in the internal program memory of the ADSP-2101. If MMAP = 1 or a boot memory address is located on any other page, the download will take place to program memory with the same addresses defined in the boot kernel.

At this point the communications program is configuring the PC as a terminal with the other device being the host; any tranfer from the PC is defined as upload. If you are using Procomm as the terminal emulation software, press (PgUp) key after selecting this entry. Then, select ASCII as the tranfer protocol. Finally, type in the full file name including the .EXE extension. Downloading from the PC will begin and a message, "Download In Progress!" , will be displayed on the screen.

Upload To Host

The contents of both program memory and data memory can be uploaded to the PC a segment at a time. Two file formats are generated: the Memory Image File (.EXE) format and the Normal format. The Memory Image file format has the same format as the format generated by the Linker. In other words, the uploaded file can be re-loaded to the emulator later on. The Normal format is in regular ASCII format which you can see the contents of the file like a regular text file. But you cannot re-load this file format back to the emulator. A sample Upload To Host is shown below.

```
================================================
|        "E Z - I C E"     ADSP-2101   EMULATOR         |
|                  Basic Command Menu                  |
|                                                      |
|        Read/Write Registers                          |
|        Read Data Memory                              |
|        Write Data Memory                             |
|        Read Program Memory                           |
|        Write Program Memory                          |
|        Reset Processor                               |
|        Single Step                                   |
|        Set, Clear, List Breakpoints                  |
|        Run Until Breakpoint                          |
|        Boot From External EPROM                      |
|        Download From Host                            |
|        Upload To Host                                |
|        Display Stacks                                |
|                                                      |
================================================
Type 1 For PM, 2 For DM : 1
Type 1 For .EXE Format, 2 For Normal Format : 1
Enter Hex Starting Address   : 0
Enter Hex Number Of Locations To Upload : 800
Hit Carriage Return To Continue
```

If Procomm is used as the terminal emulation software, press (PgDn) key after entering the number of locations to upload and before pressing (Enter←) again. Then, Procomm provides a menu asking for the transfer protocol to use; your response must be ASCII. Procomm next request the name and extension of the file to download. Type in the filename and extension. File transfer from EZ-ICE to PC takes place. When the transfer stops press the ESC key to return Procomm to terminal mode. Next press (Enter←) to return to Basic Command Menu.

Display Stacks

The Display Stacks is the last entry on the Basic Command Menu. It allows us to view the Emulator's program counter, status, and count stacks as well as the current down counter (CNTR) value. A snapshot of the Display Stacks screen is shown below.

```
                                            CNTR
                                            0006

         PC Stack        Status Stack        Count Stack

  Top>      0104                        Top>    0400
            0526                                0800
            1A3C
```

```
      ** Hit Carriage Return To Return To Basic Command Menu **
```

Note that since any of these values cannot be changed from within this display, pressing (Enter) will return to the top of the Basic Command Mneu.

4.5 SUMMARY

In this chapter, we provided a tutorial approach to learning ADSP-2101's software and hardware development tools. After following all the examples and completing exercises contained in this chapter, the user should be able to appreciate the inner working of the microcomputer, the effect of instructions on various control registers, the use and purpose of the Simulator and the Emulator, and the integration of the ADSP-2101 in its target environment. This chapter is a prelude to interesting experiments, projects and applications that will follow in the remaining chapters.

chapter 5

LABORATORY EXPERIMENTS
USING THE ADSP-2101

5.1 INTRODUCTION

The first four chapters furnished some insight into several hardware and software aspects of the microcomputer. In this and subsequent chapters we provide laboratory experiments and projects in digital signal processing using the ADSP-2101 microcomputer. We assume that students (and readers) are either familiar with or concurrently learning the principles of digital signal processing. Therefore, we begin in this chapter with introductory experiments and programs. The remaining chapters deal with more advanced programs and projects.

This chapter is devoted to experiments which incorporate all basic operations done in DSP. They range in intricacy from the simplest sampling operation to the most elaborate waveform generators. In each experiment we introduce a new processing step as well as a new programming concept. This, we believe, will enhance the learning ability of students by concentrating on a few simple aspects at a time and by taking time to analyze the results more thoroughly. In this respect we will treat the ADSP-2101 microcomputer as an instrument which is to be learned and used effectively for understanding DSP principles and algorithms.

There are several experiments in this chapter which are organized into four sections. In Section 5.2, we begin with A/D and D/A operation and discuss the effect of aliasing. This basic operation is described using the TALKTHRU program and, since it is the first full program, it is explained in detail. The elementary DSP operations, i.e., shift, scale, and add are introduced in Section 5.3. These operations are used to simulate effects of acoustic delay and echo. Section 5.4 explains the implementation of difference equations which are necessary in any filtering operation. Experiments on convolution and recursive filtering are based on this implementation. Finally, Section 5.5 describes how to implement transcendental functions and random number generators on the ADSP-2101. These generators along with the Timer unit of the microcomputer are used in experiments to generate and display waveforms. All these concepts are described using simple programs, while in experiments students are asked to write more elaborate programs. The program listings (.DSP files) given in this chapter and their executables (.EXE files) are on the diskette which is available from Analog Devices, Inc. (see Preface for details).

Before beginning the actual work, the user should check to make sure that the EZ-ICE and the EZ-LAB are successfully set up, tested and connected to a PC. For audio input/output, a microphone and a speaker should be connected to the EZ-LAB according to its manual [4]. Similarly for DAC outputs of the EZ-LAB, an oscilloscope should be available. Finally, the user should verify that the cross-software is properly installed on the PC and that a communication software to emulate VT100 is available.

The programs described in this chapter use one of the two architecture description files, referred to as EZLAB1.ACH and EZLAB2.ACH. These files are also available on the diskette. In most programs, the internal program and data memory available in the microcomputer is sufficient. Hence the default target system is described in EZLAB1.ACH file. However in the DELAY and ECHO programs described in section 5.3, external memory located on the EZ-ICE is used to generate delays up to one second. This configuration is described in EZLAB2.ACH file. It is not necessary to use these files and they are given for reference purposes only. In fact wherever possible, students should write their own architecture description files to incorporate their own defined ports, symbols, etc.

5.2 A/D AND D/A CONVERSION

The first step in digital signal processing is sampling and quantization of analog signals or A/D conversion. The final step in producing the processed analog signal requires reconstruction or D/A conversion. These are perhaps the simplest operations to perform. Therefore in the first experiment, we will consider these operations and their effects on analog signals.

The purpose of this experiment is two-fold. First, in almost all applications, we will need code to input and output signal data to and from the processor. Thus the program code from this section can be used several times in other experiments, projects, or programs. Second, a simple A/D and D/A program, called *Talk Through Program* can be used to test our hardware setup in the beginning of each laboratory session. Although it is not a substitute for a hardware diagnostic program, it nevertheless does provide an assurance when we are debugging more complex programs.

5.2.1 A Talk Through Program

In this program, an audio signal from a microphone is passed to a speaker through the ADSP-2101 system which acts as a link (albeit expensive). It is available on the diskette as a TALKTH-RU.DSP file. We will explain the workings of this program and also provide details about its execution on the Emulator. As Experiment 1, we will assemble, link and execute variations of this program.

The program TALKTHRU.DSP is shown in Listing 5-1. It contains assembler directives, helpful comments and instructions. Using comments and instructions (as discussed in Chapters 2 and 4) it should be possible to understand this program. However, we will discuss it in detail below. Subsequent programs will not be described in as much detail except for any new features.

```
{ ADSP-2101 Talk Through Program                              TALKTHRU.DSP

    This program takes an input sample from serial port 0 receive register
    and outputs it to serial port 0 transmit register. It is intended for
    speech signals. The serial clock is internally generated and 12.288 MHz
    processor rate gives 8 KHz sampling rate. This program can be used for
    testing the hardware.

    This program is written for EZ-ICE and EZ-LAB system with EZLAB1.ACH
    architecture file.  Assemble using ASM21.EXE and link using LD21.EXE to
    produce TALKTHRU.EXE.  Load TALKTHRU.EXE in EZ-ICE and execute.
}
.MODULE/RAM/BOOT=0/ABS=0 TALKTHRU;      {Beginning of TALKTHRU Program}
{----Interrupt Vectors---------------}
        JUMP start; NOP; NOP; NOP;      {Start Interrupt}
        RTI; NOP; NOP; NOP;             {External Pin Interrupt IRQ2}
        RTI; NOP; NOP; NOP;             {SPORT0 Transmit Interrupt}
        JUMP sample; NOP; NOP; NOP;     {SPORT0 Receive Interrupt}
        RTI; NOP; NOP; NOP;             {SPORT1 Transmit Interrupt}
        RTI; NOP; NOP; NOP;             {SPORT1 Receive Interrupt}
        RTI; NOP; NOP; NOP;             {TIMER Interrupt}
{----Initializations-----------------}
start:  AX0=0x1000;                     {SPORT0 enabled, PM Wait State 0,}
        DM(0x3FFF)=AX0;                 {BOOT Wait state 0, BOOT page 0 }
        TOGGLE FLAG_OUT;                {Lights FLAG LED}
        AX0=0x0000;
        DM(0x3FFE)=AX0;                 {All DM Wait States 0}
        DM(0x3FFB)=AX0;                 {TIMER                          }
        DM(0x3FFC)=AX0;                 {         not used,             }
        DM(0x3FFD)=AX0;                 {                      cleared}
        DM(0x3FE9)=AX0;                 {Receive                        }
        DM(0x3FFA)=AX0;                 {              Multichannels}
        DM(0x3FF7)=AX0;                 {Transmit                       }
        DM(0x3FF8)=AX0;                 {              Multichannels}
        AX0=0x6B27;                     {Multichannel disabled}
        DM(0x3FF6)=AX0;                 {Int. gen serial clock}
                                        {Receive frame sync reqd, width 0}
                                        {Transmit frame sync reqd, width0}
                                        {Int trans, receive frame sync ena}
                                        {µ-law companding, 8 bit word len}
        AX0=0x0002;
        DM(0x3FF5)=AX0;                 {Generate 2.048 MHz serial clock}
        AX0=255;
        DM(0x3FF4)=AX0;                 {Divide by 256 for 8KHz samp rate}
        AX0=0x0000;
        DM(0x3FF3)=AX0;                 {SPORT0 AUTOBUFF disabled}
        DM(0x3FF2)=AX0;                 {SPORT1 CNTL disabled}
        DM(0x3FF1)=AX0;                 {SPORT1 timing not used}
        DM(0x3FF0)=AX0;                 {SPORT1 timing not used}
        DM(0x3FEF)=AX0;                 {SPORT1 AUTOBUFF DISABLED}
        ICNTL=0x07;                     {Enable edge sensitive interrupt}
        IMASK=0x08;                     {Enable SPORT0 Interrupt}
{----Wait for sample-----------------}
wait:   IDLE;                           {Wait until next sample appears at}
        JUMP wait;                      {                          SPORT0}
{----Process sample------------------}
sample: AX0=RX0;                        {Put received sample in AX0}
        TX0=AX0;                        {Transmit sample value in AX0}
        RTI;                            {Return from Interrupt}
.ENDMOD;                                {End of TALKTHRU Program}
```

———————————— Listing 5-1: Talk Through Program ————————————

The program begins with a comment field containing a brief summary and information for its execution. The first non-comment statement is the .MODULE directive signaling the beginning of the program at absolute address 0 and boot page 0. The next seven lines contain interrupt service vectors, each four instructions long. The first of these is the start interrupt which transfers program control to *start:* label. Since SPORT0 provides our I/O, only the fourth vector provides the necessary servicing (which in this case is directed towards *sample:* label). Therefore the remaining interrupt vectors provide no servicing. This block of vectors is a required feature and must be used in all programs. Depending on each program's I/O, the way each interrupt is serviced will change accordingly.

After the execution begins, the program jumps to *start:*. Here for the next thirty lines, the data memory mapped control registers are setup. In this first program, we explicitly initialized all control registers. According to the *ADSP-2101 User's Manual* [1], the registers have default values at startup and only few registers are required to be setup. However, it is a good programming practice to initialize all control registers and avoid any possible problems later on. For a complete explanation on the initialization of these registers, refer to Appendix D of [1].

The first of these is the System Control Register at DM location 0x3FFF. It is initialized so that SPORT0 is enabled (bit 12), FI/FO pins are enabled and PM wait state, BOOT wait state and BOOT page are set to 0. The next instruction toggles the FLAG OUT pin so that the corresponding LED lights up on the EZ-LAB board. Since the Emulator lacks trace capability, this instruction can be used to debug programs by suitably placing it in a suspect part of the code. By looking at the status of this LED, we can determine which part of the program is being executed. Otherwise, it does not contribute to the working of the program. Control registers from DM locations 0x3FF7 to 0x3FFE are all reset to 0. Therefore, all DM wait states are set to 0, and TIMER and multichannel operations are disabled.

Since SPORT0 is enabled, its Control Register located at 0x3FF6 is an important one. In this register bits 0-2, 5, 8, 9, 11, 13 and 14 are set to 1. This selects μ-law companding, 8-bit word length, and transmit/receive frame sync. Also the serial clock is internally generated. Note that bits 0-3 denote serial word length parameter SLEN and the word length is equal to SLEN + 1. Therefore, SLEN is set to 7. Most of the values in this register will be same for many programs in this laboratory.

The next two registers are important in setting the sampling rate for our operation. The SPORT0 Serial Clock Divide Modulus (SCLKDIV) register at 0x3FF5 generates the serial clock frequency (SCLK) according to the following formula:

$$SCLK \; = \; \frac{CLKOUT}{2(SCLKDIV + 1)}$$

where CLKOUT is the processor frequency. In our lab setup, it is assumed that the ADSP-2101 is operating at 12.288 MHz. Therefore, SCLKDIV = 2 provides a serial clock frequency of 2.048 Mhz. The sampling frequency is derived from SCLK using the value in the RFSDIV register located at 0x3FF4. It is given by,

$$Sampling \; Frequency \; = \; \frac{SCLK}{RFSDIV + 1}$$

Hence a value of 255 in RFSDIV gives a sampling frequency of 8KHz, suitable for speech signals up to 4KHz in bandwidth. From the above two formulas, it can be seen that the sampling frequency depends on two parameters, SCLKDIV and RFSDIV. This also means that a wide range of sampling frequencies can be generated and that there is no unique way of determining a sampling frequency.

The next five memory mapped control registers are reset to zero since their functions are not required in this program. The control registers ICNTL and IMASK, which are not memory mapped, are also important in every program. The ICNTL register, which is 5-bits wide and which controls the interrupt sensitivity, is set to 7 so that all interrupts are edge sensitive. Finally the IMASK register which is 6-bits wide is set to 8 so that SPORT0 receive interrupt is enabled. It is a good programming practice to set this register, after all other parameters of the serial ports are initialized and before interrupts are allowed, to prevent the processor from vectoring to an interrupt routine before the set-up is complete. Note that the ICNTL register should be set before the IMASK because as soon as the IMASK is set, interrupts are accepted and the processor could vector before the ICNTL is executed.

So far in this program, we spent almost thirty lines to setup the processor. It seems a lot for this simple A/D — D/A conversion program. However note that we covered all control registers even though not every one of them is required, so that we will be exposed to them and be aware of them for any future use. Now the program is ready to process input samples. The processor loops on the IDLE instruction until the interrupt from SPORT0 is received. The program is thus interrupt-driven. When the interrupt occurs, program control shifts to the service routine by first jumping to location 0x000C (*jump sample*) and then jumping to location *sample:*.

All further activity takes place in the interrupt service routine. The input sample which is in the RX0 register of SPORT0 is moved to the AX0 register. Note that any internal data register can be used. In the next cycle, the input sample is moved from AX0 to TX0, the transmit register of SPORT0. After the return from interrupt, execution resumes at the *wait:* loop. The last statement .ENDMOD signals the end of the program.

Now we are ready to execute this program and at the same time check out the EZ-ICE and EZ-LAB hardware. Copy TALKTHRU.DSP from the diskette to your working directory. Before beginning this demonstration, make sure that all systems are properly powered up and running. The EZ-ICE should be configured according to the architecture description file EZLAB1.ACH with overlay memory disabled.

The first task is to assemble the program. At a DOS prompt, issue the following command.

```
asm21 talkthru
```

This file should assemble properly without any errors. The assembler creates three new files with extensions .OBJ, .CDE, .INT which are used by other cross-software programs. Now create the binary executable file .EXE by linking the object code. Issue the following command at the DOS prompt.

```
ld21 talkthru -a ezlab1 -e talkthru
```

Once again the linker should produce TALKTHRU.EXE without any error. This file can now be loaded into the EZ-ICE. Start the terminal emulation program and activate the Emulator. This is discussed in Chapter 4. The Emulator will respond with its command menu. Before loading the program, it is a good practice to reset the processor. From the command menu select *Reset Processor* option. Then select *Download From Host* option and load the TALKTH-RU.EXE file. Our first program is now ready to control the ADSP-2101.

Run the program by selecting *Run Until Breakpoint*. A "Processor Running" message will appear in the command menu window. If the entire system is functioning properly, an audible click will be heard from the speaker. The microphone is now connected to the speaker through the processor and any speech spoken into the microphone will be heard from the speaker. By pressing any key on the keyboard the processor is halted, thereby breaking the connection. Try several times repeating this process.

Through this simple talk through program we covered most of the aspects of ADSP-2101 processor's programming. Any new features will be dealt in similar detail as we go along. For Experiment 1A, we will once again consider the same A/D — D/A conversion but with emphasis on indirect addressing. In Experiment 1B, we will study the aliasing phenomenon which is the effect of sampling.

EXPERIMENT 1A

In Listing 5-1, all memory mapped control registers were initialized using the direct addressing mode. In this mode, the DM location was addressed directly using an hexadecimal address and was assigned a value from one of the data registers (in this case AX0). This portion of the program is not completely legible because an hexadecimal address of the control register does not correlate with the function of that register. However, It is possible to give symbolic names to these addresses through the .CONST directive of the Assembler. For example, consider the directive

```
.CONST sys_cont_reg=0x3FFF;
```

in which the variable *sys_cont_reg* (a symbolic name for the system control register) is set to the required address of the system control register. Now in the initialization portion of our program we can use the following instruction:

```
DM(sys_cont_reg)=AX0;
```

instead of

```
DM(0x3FFF)=AX0;
```

This style of programming will make all our programs readable especially when there is a need for debugging. Furthermore, it is not necessary to actually insert these .CONST directives in the main program. We can create a header file, say *const.h*, containing all the necessary .CONST directives and then include it in the main program using the directive

```
.INCLUDE <const.h>;
```

This directive keeps the main program short and avoids duplicating .CONST directives in every program which needs them.

The ADSP-2101, at reset, initializes all necessary control registers so that serial ports, autobuffering, timer, and interrupts are disabled. Therefore, write and execute a new program EXPMT-1A.DSP containing the initialization approach using symbolic names for those control registers which were required in the Talk Through program. Follow the procedure shown below:

1. Provide symbolic names for the control registers mapped at the locations 0x3FFF, 0x3FFE, 0x3FF6, 0x3FF5, and 0x3FF4.
2. Create a text file, *const.h*, containing the appropriate .CONST directives for the above locations.
3. Modify the TALKTHRU.DSP program by inserting a .INCLUDE directive and rewriting the initialization code using the symbolic names created in the *const.h* file.
3. Save the new program as EXPMT-1A.DSP.
4. Assemble, link and execute this program in the Emulator.
5. Verify that this talk through program is also working properly.

There is another approach using indirect addressing for initializing memory-mapped control registers of the ADSP-2101. In this approach, the DM location is accessed via Index and Modify registers in the DAG but a data value can be directly transferred to any DM location. This saves one cycle of execution. However it is not a recommended approach. If DAG registers are incorrectly initialized or mistakenly overwritten, the control registers will be incorrectly set. This type of error can be difficult to detect in programs.

5.2.2 Studying the Effect of Aliasing

From the theory of sampling, we know that if an analog signal is sampled at F_s samples per second then the resulting discrete signal has frequencies up to $F_s/2$ Hz. Any frequencies above $F_s/2$ Hz are aliased into lower frequencies. We will verify this fact using our ADSP-2101 system. In order to "see" aliasing, we will have to connect an oscilloscope to the output port. In this section, we will learn how to use the memory mapped D/A converter (DAC) of the target system EZ-ICE for this purpose. We will also change the sampling frequency to 4 KHz so that all frequencies above 2 KHz will be aliased. We will call this program ALIASING.DSP and is shown in Listing 5-2.

The main difference between this and the previous program is the use of DAC port of the EZ-LAB. (The architecture description file EZLAB1.ACH contains the port addresses.) The two *.PORT* directives declare memory mapped I/O ports in the DM at 0x1000 (*WRITE_DAC0*) and 0x2000 (*LOAD_DAC*, which is used to update DAC outputs). Because of DAC setup timing requirements, two wait states are required when writing to the DAC. Therefore DM Wait State Control Register at 0x3FFE is set so that its DWAIT2 field is equal to 2, which makes the register value equal to 0x0080. Finally, two extra lines of code are needed in the SPORT0 receive service routine to load and transfer data to DAC0.

Copy ALIASING.DSP from the diskette to your working directory. Then assemble and link to create ALIASING.EXE using the procedure discussed in the previous section. Connect

```
{ ADSP-2101 Program to demonstrate Aliasing                      ALIASING.DSP

    This program is intended to demonstrate aliasing effects. It takes an
    input sample from serial port 0 receive register and outputs it to
    serial port 0 transmit register. In addition, the output sample is also
    available at DAC0 of EZ-LAB to which an oscilloscope can be connected.
    The serial clock is internally generated and the sampling rate is 4
    KHz.

    This program is written for EZ-ICE and EZ-LAB system with EZLAB1.ACH
    architecture file.  Assemble using ASM21.EXE and link using LD21.EXE to
    produce ALIASING.EXE.  Load ALIASING.EXE in EZ-ICE and execute.
}
.MODULE/RAM/ABS=0 ALIASING;                 {Beginning of ALIASING Program}
.PORT WRITE_DAC0;                           {Memory Mapped DAC port at 0x1000}
.PORT LOAD_DAC;                             {Memory Mapped port at 0x2000}
{----Interrupt Vectors----------------}
        JUMP start; NOP; NOP; NOP;          {Start Interrupt}
        RTI; NOP; NOP; NOP;                 {External Pin Interrupt IRQ2}
        RTI; NOP; NOP; NOP;                 {SPORT0 Transmit Interrupt}
        JUMP sample; NOP; NOP; NOP;         {SPORT0 Receive Interrupt}
        RTI; NOP; NOP; NOP;                 {SPORT1 Transmit Interrupt}
        RTI; NOP; NOP; NOP;                 {SPORT0 Receive Interrupt}
        RTI; NOP; NOP; NOP;                 {TIMER Interrupt}
{----Initializations : similar to TALKTHRU program}
start:  AX0=0x1000;
        DM(0x3FFF)=AX0;                     {SPORT0 enabled}
        AX0=0x0080;
        DM(0x3FFE)=AX0;                     {DWAIT2 = 2}
        AX0=0X0000;
        DM(0x3FFB)=AX0;                     {TIMER not used , cleared}
        DM(0x3FFC)=AX0;
        DM(0x3FFD)=AX0;
        DM(0x3FE9)=AX0;                     {Receive Multichannels}
        DM(0x3FFA)=AX0;
        DM(0x3FF7)=AX0;                     {Transmit Multichannels}
        DM(0x3FF8)=AX0;
        AX0=0x6B27;                         {Multichannel disabled}
        DM(0x3FF6)=AX0;                     {Int. gen serial clock}
        AX0=0x0002;
        DM(0x3FF5)=AX0;                     {Generate 2.048 MHz serial clock}
        AX0=511;
        DM(0x3FF4)=AX0;                     {Divide by 512 for 4KHz rate}
        AX0=0x0000;
        DM(0x3FF3)=AX0;                     {SPORT0 AUTOBUFF disabled}
        DM(0x3FF2)=AX0;                     {SPORT1 CNTL disabled}
        DM(0x3FF1)=AX0;                     {SPORT1 timing not used}
        DM(0x3FF0)=AX0;                     {SPORT1 timing not used}
        DM(0x3FEF)=AX0;                     {SPORT1 AUTOBUFF DISABLED}
        ICNTL=0x07;                         {Enable edge sensitive interrupt}
        IMASK=0x08;                         {Enable SPORT0 Interrupt}
{----wait for sample------------------}
wait:   IDLE;                               {Wait until next sample appears at}
        JUMP wait;                          {                        SPORT0}
{----process sample------------------}
sample: AX0=RX0;                            {Put received sample in AX0}
        TX0=AX0;                            {Transmit sample value in AX0}
        DM(WRITE_DAC0)=AX0;                 {Load DAC0 register}
        DM(LOAD_DAC)=AX0;                   {Transfer data}
        RTI;                                {Return from Interrupt}
.ENDMOD;                                    {End of ALIASING Program}
```

──────────────── Listing 5-2: Program to demonstrate Aliasing ────────────────

an oscilloscope to the DAC0 terminals of the EZ-LAB. Activate the Emulator, load ALIA-SING.EXE and execute the program. The ADSP-2101 processor is now available to verify aliasing effects. Speak into the microphone and listen to the output signal. Also observe the signal on the oscilloscope. Can you characterize this output speech signal?

Experiment 1B

In this experiment, we will connect various signal sources and study the effect of aliasing on them. We will also learn how to change the sampling frequency in the Emulator without changing the original program.

1. Edit EXPMT-1A.DSP program and change the sampling frequency to 4 KHz. Call this EXPMT-1B.DSP program. Assemble, link and load EXPMT-1B.EXE in the Emulator.

2. Connect a signal generator to the microphone port (disconnect microphone). Test your program for sinusoidal waveforms from 100 Hz to 4 KHz. Record your observations.

3. Select a square wave input with fundamental frequency of 1 KHz and record your observations. Repeat with fundamental frequency of 1.5 KHz. Can you justify these observations?

4. We want to change the sampling frequency from 4 KHz to 6 KHz. One way to do this is to edit EXPMT-1B.DSP, make the necessary change and then assemble, link and execute the program. There is another approach. We can make a direct change in the program memory from the Emulator's command menu without going through edit-assemble-link process. Halt the processor by pressing any key and select *Read Program Memory* option. From disassembled contents, locate the instruction, AX0=511, which is used to set sampling frequency to 4 KHz. We want to change this instruction to AX0=341 (why?). From the command menu, select *Write Program Memory* option. Enter the proper hexadecimal address in program memory at which the above change is to take place. The Emulator requests for the content of the program memory value. Unfortunately, this content must be a six-digit hexadecimal value which should be the opcode for AX0=341. Refer to Cross-Software Manual [2] to determine this opcode. After making this change we have a new A/D — D/A connector operating at 6 KHz rate. Verify this sampling frequency using sinusoidal signals.

5.2.3 Exercises

1. In the Talkthru program, incorporate DAC0 output and display the output speech signal on an oscilloscope.

2. The audio circuitry of the EZ-LAB contains a codec to which a microphone and speaker are connected. This codec is designed to operate at an 8 KHz sampling rate. Therefore our attempt to change the sampling frequency in this section may result in unpredictable behavior of the codec. (Does your observation so far confirm this?). Hence we will employ a different technique to reduce the sampling frequency while still operating the codec at an 8 KHz rate. Edit EXPMT-1B.DSP and change the sampling frequency back to 8 KHz. Modify SPORT0 Receive Interrupt service

routine so that every other sample is transmitted to the codec and to the DAC0. This is called subsampling, or down sampling, and will result in 4 KHz sampling frequency. Verify the operation of this new program using speech as well as sinusoidal signals.

3.   Write a program so that every other speech sample from the microphone is sign reversed before it is transmitted to the speaker through a SPORT0 register. This can be accomplished by using a register which toggles between 1 and -1 along with a conditional IF statement which checks the sign of this register before sign reversing the input sample. Mathematically, this operation multiples the input sequence by {1, -1, 1, -1, ...} sequence. It modulates signal from a base band of 0 Hz to 4 KHz which causes frequency inversion due to aliasing. This is known as *speech scrambling through frequency inversion*. Characterize the frequency response of the output sequence in terms of that of the input sequence and justify the name frequency inversion. Speak into the microphone and record your observations regarding the scrambled speech.

5.3 ELEMENTARY DSP OPERATIONS

After getting the ADSP-2101 processor to perform the simplest operation of A/D and D/A conversion, we are now ready to program three elementary DSP operations. These are: *delay* (data move), *scale* (constant multiplier), and *add* operations. The delay operation provides a shift of one to several sample-intervals or even shifts of a few seconds. In the *Delay* section, we will simulate an acoustic delay-line by delaying the input signal up to 1 second (maximum possible in EZ-ICE). The constant multiplier scales a sample while the adder adds two signal samples to form another sample. In fact almost all DSP algorithms can be done using these three elementary operations and the ADSP-2101 processor is optimized for these operations. In the *Echo* section, we will simulate an acoustic echo by adding an input sample to its delayed version and suitably scaling the sum to avoid overflow in the processor. In addition to their learning value, these programs also provide interesting audio effects.

The single most important programming concept in this experiment is the circular buffer. The incoming data is stored in the buffer whose length is equal to the total amount of delay (in number of samples) we want to generate. The stored value in a buffer location is first transmitted to the serial port and then the incoming sample value is stored at that buffer location. The buffer location is advanced by one and the process continues. Due to the circular nature of the buffer, the stored sample is read out after one rotation (equal to the length) of the buffer thus providing the necessary delay. The addressing of the circular buffer is provided by the index, modify and length registers.

In the last experiment, we executed programs on the Emulator. As a learning experience, we will execute the *Delay* and *Echo* programs in this experiment from the Simulator. Obviously the microphone and the speaker cannot be interfaced with the Simulator. However their actions can still be simulated through disk files. This requires using the Simulator commands to open data files as serial ports and also to simulate interrupts. Once we are satisfied by verification on the Simulator that these programs are working properly, then we will execute them on the Emulator and demonstrate their audio effects. This is the framework of our Experiment-2.

5.3.1 A Delay Program

In this program, we will demonstrate the delay operation using the Simulator. If $x(n)$ is the input sample and $y(n)$ is the output sample then the operation performed by this program is

$$y(n) = x(n-N)$$

where N is the amount of delay in samples. Since the program is executed in the Simulator, there are some particular features which we will explain in detail. The program DELAY.DSP is shown in Listing 5-3.

```
{ ADSP-2101 Delay Program                                         DELAY.DSP

     This program takes an input sample from memory-mapped port_in at loca-
     tion 0x1000 and outputs it to another memory-mapped port_out at loca-
     tion 0x1001 after a delay of 10 samples. It is intended for the
     Simulator. The interrupt used is IRQ2, which must be simulated in the
     Simulator. An appropriate architecture file must also be written to
     include memory mapped I/O ports.

     Assemble this program using ASM21.EXE and link using LD21.EXE to pro-
     duce DELAY.EXE.  Load DELAY.EXE in the Simulator, open file INPUT.DAT
     at 0x1000 as input and OUTPUT.DAT at 0x1001 as outout. Simulate IRQ2
     interrupt and then execute.
}
.MODULE/RAM/ABS=0 DELAY_SIM;                {Beginning of DELAY Program}
.PORT     port_in;                          {Memory mapped Input Port}
.PORT     port_out;                         {Memory mapped Output Port}
.CONST    delay_size=10;
.VAR/DM/CIRC cir_buff[delay_size]; {Circular buffer, length delay_size}
{----Interrupt Vectors----------------}
          JUMP start; NOP; NOP; NOP;        {Start Interrupt}
          JUMP delay; NOP; NOP; NOP;        {External Pin Interrupt IRQ2}
          RTI; NOP; NOP; NOP;               {SPORT0 Transmit Interrupt}
          RTI; NOP; NOP; NOP;               {SPORT0 Receive Interrupt}
          RTI; NOP; NOP; NOP;               {SPORT1 Transmit Interrupt}
          RTI; NOP; NOP; NOP;               {SPORT0 Receive Interrupt}
          RTI; NOP; NOP; NOP;               {TIMER Interrupt}
{----Initialization: Circular Buffer----}
start:    I0=^cir_buff;                     {Index to top of cir_buf}
          L0=%cir_buff;                     {Enable circular buffer}
          M0=0;                             {No post-increment}
          M1=1;                             {Post-increment by 1}
{----Clear circular buffer-------------}
          CNTR=delay_size;                  {Set up counter}
          DO clr_buf UNTIL CE;              {Initialize all cir_buff values}
clr_buf: DM(I0,M1)=0;                       { to 0}
          ICNTL=0x07;                       {Enable edge sensitive interrupt}
          IMASK=0X20;                       {Enable IRQ2 interrupt}
{----Wait for Sample------------------}
wait:     IDLE;                             {Wait until next sample}
          JUMP wait;
{----Process sample and generate delay----}
delay :   AX0=DM(I0,M0);                    {Get delayed sample from cir_buf}
          DM(port_out)=AX0;                 {send delayed sample out}
          AX0=DM(port_in);                  {Get input sample in AX0}
          DM(I0,M1)=AX0;                    {Move current sample to cir_buf}
          RTI;                              {Return from Interrupt}
{----End of program-------------------}
.ENDMOD;                                    {End of DELAY Program}
```

——————————— Listing 5-3: Delay Program ———————————

The I/O operation for this program is simulated through disk files since the Simulator cannot interface with any real physical device. The Simulator is a piece of software and has no concept of time. Therefore, although it is possible to connect SPORT0 receive and transmit registers to these files, operation at any sampling rate makes no sense in this case. Use of memory mapped I/O ports is sufficient to interface with this program (or any program written for the Simulator). Then one has to write an architecture description file to describe these I/O ports. Finally, one has to simulate the receipt of input samples as an interrupt. This can be done through the external interrupt IRQ2. Now all the elements of this program are in place.

The program begins with assembler directives. Memory mapped input is described as *port_in* which is mapped at DM address 0x1000 (this is arbitrary) while the output is described as *port_out* mapped at DM location 0x1001. It is important that these values be specified in the architecture description file. The .CONST directive assigns the value 10 to the variable *delay_size* which then can be used in the program. The .VAR directive declares *cir_buff* as a circular buffer of size *delay_size*.

In the interrupt vectors segment, only start and IRQ2 interrupt service routines are required. Since SPORTs, Timer, and external memory features are not used in this program, control register initializations are not required.

The circular buffer is addressed via indirect addressing using DAG0. The first instruction in the initialization segment loads the index register with the address of the first memory location of the circular buffer *cir_buff*. This location is actually determined by the linker. Therefore "^" operator is used as a pointer to this address. Similarly "%" operator is used to indicate the length of the buffer which is assigned to the length register L0. M0 and M1 are modify registers which are set to the appropriate values as explained below.

Next, the circular buffer is initialized to zero (or cleared) using the counter CNTR and a DO ... UNTIL CE loop. The counter is set to the length of the buffer and the DO LOOP is executed until this counter expires. This completes the initialization of the DAG registers and circular buffer. Now interrupt sensitivity is set using the ICNTL register and the IRQ2 interrupt is enabled with the IMASK register. The program is ready to process input samples. The processor loops on the IDLE instruction until the interrupt is received from IRQ2. When this interrupt occurs, the program control shifts to *delay:* location.

All further activities take place in the interrupt service routine. First the stored sample from the circular buffer is transmitted to the output port via the AX0 register. Since M0 is equal to 0, I0 is not advanced and it still points to the same address location from which the sample is sent out. Then the sample from the input port is read into this location and I0 is advanced by one since M1 is equal to 1. This sample will be read out after one rotation of the circular buffer thus generating the required delay.

Now we are ready to execute this program. Copy DELAY.DSP from the diskette to your working directory. Create an appropriate DELAY.SYS file and generate its DELAY.ACH file. Finally, assemble and link the Delay program to generate DELAY.EXE file. Do not forget to generate a .SYM file which is required in the Simulator.

We are now ready to execute this program in the Simulator. First we need input samples in a disk file. Create an ASCII INPUT.DAT file containing a few (at least twice the number of samples being delayed) sample values, with one value per line. Invoke the Simulator and open program and data memory windows. Execute the following commands from the command window:

```
cr
l 'delay'
o dm[0x1000] < 'input.dat'
o dm[0x1001] > 'output.dat'
i 2 10
g
```

The first command resets the processor. The second command loads the delay program. The next two commands open disk files as memory mapped I/O ports at respective addresses. The fourth command simulates interrupt IRQ2. Since timing is not important and since the interrupt routines takes five cycles, the interrupt is simulated after every ten instruction cycles. Finally, the last command executes the program. The program halts whenever the INPUT.DAT file runs out of sample values. Observe the program and data memory windows during the execution. After the program stops, exit from the Simulator and check the OUTPUT.DAT file to verify the operation of the Delay program.

5.3.2 Overlay Memory

The ADSP-2101 microcomputer has 1K 16-bit words of data memory which can be used to store samples in a circular buffer. This can simulate delays up to 128 miliseconds at an 8 KHz sampling rate. To generate delays up to 1 sec, we need a data memory of size 8K which then must be an external memory. The EZ-ICE probe has 8K 24-bit words of high speed static memory available as overlay memory. We will use this memory as data memory. It replaces selected portions of the target system's (EZ-LAB) memory. Use of the overlay memory is firmware controlled and is initially disabled upon probe reset.

The position of jumper JP1 on the probe is used to select the segment of ADSP-2101 memory to be replaced by overlay memory. JP1 is located next to the RESET switch on the top of the probe. Jumper position 2 selects the overlay memory as data memory. This configuration still has to be enabled and the actual enabling of the overlay memory is software selectable from the initial menu, once an interface is established with the PC.

Overlay memory begins at address 0x0000 and continues through address 0x1FFF. Since EZ-LAb's D/A converters are memory mapped between 0x1000 and 0x1003, they are no longer available once overlay memory is enabled as data memory. This changes the environment of the target system and necessitates writing another appropriate system builder file to reflect external data memory beginning at absolute address 0.

Before beginning this experiment, select position 2 of jumper JP1 to use the overlay memory. When you activate the Emulator through the terminal emulation program, do not forget to enable the overlay memory from the initial menu. Finally, generate an architecture description file EZLAB2.ACH using the System Builder. In case of difficulties, file EZLAB2.SYS is available on the diskette.

EXPERIMENT 2A

Using the overlay memory, we will simulate a delay in the samples of up to 1 second in the Emulator. Also we will obtain input from the microphone and output samples to the speaker of the EZ-LAB.

1. Modify the Delay program by incorporating code from the Talkthru program to simulate delays up to 1 second in the system operating at an 8 KHz sampling rate. (Circular buffer length must be 8000). Save this program as EXPMT-2A.DSP.

2. Simulate delays of 10 ms, 100 ms, 500 ms and 1 sec. Demonstrate your program and record your observations.

5.3.3 An Echo Program

In a concert hall live performance, we hear sound in different stages. The direct sound is followed by multiple reflections or echoes. We will try to simulate this situation using our ADSP-2101 system. Using one delayed reflection, the output sample $y(n)$ is given by

$$y(n) = \alpha x(n) + \beta x(n-N); \qquad \alpha + \beta \le 1$$

Using our Delay program it is easy to do this operation. We have to add a scaled delayed sample to the scaled current sample with proper care to avoid overflow. Listing 5-4 shows an Echo program written for the Simulator.

```
{ ADSP-2101 Echo Program                                       ECHO.DSP

    This program takes an input sample from memory-mapped port_in at loca-
    tion 0x1000. It outputs to another memory-mapped port_out at location
    0x1001 the average of the current sample and a sample delayed by 10
    samples. It is intended for the Simulator. The interrupt used is IRQ2,
    which must be simulated in the Simulator. An appropriate architecture
    file must also be written to include memory mapped I/O ports.

    Assemble this program using ASM21.EXE and link using LD21.EXE to pro-
    duce ECHO.EXE.  Load ECHO.EXE in the Simulator, open file INPUT.DAT at
    0x1000 as input and OUTPUT.DAT at 0x1001 as outout. Simulate IRQ2
    interrupt and then execute.
}
.MODULE/RAM/ABS=0 ECHO_SIM;                 {Beginning of ECHO Program}
.PORT     port_in;                          {Memory mapped Input Port}
.PORT     port_out;                         {Memory mapped Output Port}
.CONST    delay_size=10;
.VAR/DM/CIRC cir_buff[delay_size]; {Circular buffer, length delay_size}
{----Interrupt Vectors----------------}
          JUMP start; NOP; NOP; NOP;        {Start Interrupt}
          JUMP delay; NOP; NOP; NOP;        {External Pin Interrupt IRQ2}
          RTI; NOP; NOP; NOP;               {SPORT0 Transmit Interrupt}
          RTI; NOP; NOP; NOP;               {SPORT0 Receive Interrupt}
          RTI; NOP; NOP; NOP;               {SPORT1 Transmit Interrupt}
          RTI; NOP; NOP; NOP;               {SPORT0 Receive Interrupt}
          RTI; NOP; NOP; NOP;               {TIMER Interrupt}
{----Initialization: Circular Buffer----}
start:    I0=^cir_buff;                     {Index to top of cir_buf}
          L0=%cir_buff;                     {Enable circular buffer}
          M0=0;                             {No post-increment}
          M1=1;                             {Post-increment by 1}
{----Clear circular buffer-------------}
          CNTR=delay_size;                  {Set up counter}
          DO clr_buf UNTIL CE;              {Initialize all cir_buff values}
clr_buf: DM(I0,M1)=0;                       { to 0}
```

```
           IMASK=0X20;                              {Enable IRQ2 interrupt}
{----Wait for Sample------------------}
wait:      IDLE;                                    {Wait until next sample}
           JUMP wait;
{----Process sample and generate delay----}
delay :    AY0=DM(I0,M0);                           {Get delayed sample from cir_buf}
           SI=DM(port_in);                          {Get input sample in SI}
           SR=ASHIFT SI by -1 (HI);                 {Scale down current sample by 50%}
           AR=SR1+AY0;                              {Sum current and delayed samples}
           DM(port_out)=AR;                         {send delayed sample out}
           DM(I0,MI)=SR1;                           {Move current sample to cir_buf}
           RTI;                                     {Return from Interrupt}
{----End of program------------------}
.ENDMOD;                                            {End of ECHO Program}
```
———————————— **Listing 5-4: Delay Program** ————————————

The only difference between this and the previous program is in the interrupt service routine. The input sample is shifted to the right by one bit to obtain 50% scaling and added to the delayed sample (which was previously scaled by 50%) to produce the output sample. Execute this program in the Simulator using the same INPUT.DAT file as the source for the input samples. Check the OUTPUT.DAT file and verify the operation of this program.

EXPERIMENT 2B

In this experiment, we will simulate the concert hall effect using the Emulator. Make sure that you are using overlay memory and the corresponding .ACH file.

1. Modify the EXPMT-2A.DSP file to include portions of the Echo program. Call this EXPMT-2B.DSP.
2. Simulate echoes of duration 10 ms and 50 ms. Demonstrate your results.
3. To obtain the proper concert hall effect, we need at least two reflection delays of small duration. Modify EXPMT-2B.DSP to include one additional delay with proper scaling. Experimentally demonstrate the concert hall effect using this new program.

5.3.4 Exercises

1. Implement the frequency inversion program of Exercise 5.2.3-3 using the operations discussed in this section. In particular, use a circular buffer of length 2 containing coefficients 1 and -1.

2. Write an efficient program to implement the following input/output operation

$$y(n) \quad = \quad x(n) + \alpha x(n-N) + \alpha^2 x(n-2N)$$

5.4 DIFFERENCE EQUATION IMPLEMENTATIONS

One of the routine operations in DSP is a filtering operation. This operation is typically implemented in practice using a difference equation approach. A difference equation implementation of a filter requires shifting, scaling, and summing of input/output samples to obtain a processed sample. In the last experiment, we studied these basic operations on the ADSP-2101 microcomputer which is optimized to perform difference equation operations efficiently. In this experiment, we will study and develop more intricate programs on difference equations, both in the Simulator and on the EZ-ICE/EZ-LAB system.

Digital filters are classified as FIR (Finite-duration Impulse Response) or IIR (infinite-duration Impulse Response) filters. FIR filters are characterized as Moving Average (MA) filters and their difference equation implementation leads to the familiar linear convolution operation. In the *Convolution* section, we will first study a simulator version of the program suitable for convolving two finite duration sequences. In Experiment 3A, we will develop a more detailed program to process a real-time (and hence of infinite duration) input signal on the Emulator. IIR filters, on the other hand, are characterized as recursive or Auto Regressive (AR) filters and are implemented by a general linear constant coefficient difference equation. In the *Recursive Filter* section, we will again study a simulator version of the program which computes an impulse response of a first-order recursive filter. In Experiment 3B, we will develop a program to implement a general difference equation on the Emulator. Filtering operations on speech signals can now be demonstrated using these programs.

The single most important programming concept in this experiment is the multifunction instruction. The ADSP-2101 processor can perform multiply/add, data memory read, and program memory read operations in one instruction cycle. The internal architecture is designed around this concept for efficient DSP applications. Since the MA and AR parts of the difference equation involve multiply/add operations on sequentially stored data, one multifunction instruction placed inside a loop is sufficient to implement each part. This is the essence of our difference equation implementation.

5.4.1 A Convolution Program

In this program, we will implement an FIR filter using the linear convolution operation. If $x(n)$ is the input and $y(n)$ is the output of this filter then its difference equation representation is given by

$$y(n) = b_0 x(n) + b_1 x(n-1) + \ldots + b_{M-1} x(n-M+1)$$

$$= \sum_{m=0}^{M-1} h(m) x(n-m)$$

where $b_n, 0 \leq n \leq M-1$ are filter tap weights, $h(n)$ is the impulse response, and

$$h(n) = \begin{cases} b_n & , \quad 0 \leq n \leq M-1 \\ 0 & , \quad\quad else \end{cases}$$

If $x(n)$ is an N-point sequence then $y(n)$ will be a $N+M-1$-point sequence after filtering is completed. We will demonstrate this using the Simulator. The appropriate program CONVOLV.DSP is shown in Listing 5-5.

In this program, there are no input/output operations. Both $x(n)$ and $h(n)$ are loaded in the data memory from disk files, while $y(n)$ is available only in data memory. Since this program is intended for short duration sequences and since we have studied how to handle input/output in the Simulator, these operations are not simulated in this program.

```
{ ADSP-2101 Convolution Program                              CONVOLV.DSP

    This program performs a linear convolution of M-point sequence 'h(n)' and
    N-point sequence 'x(n)'.  It is intended for the Simulator.  Sequence
    'h(n)' is loaded from disk file 'H(N).DAT' while sequence 'x(n)' is loaded
    from file 'X(N).DAT'.   The output sequence 'y' is stored in some data
    memory locations. Architecture file EZLAB1.ACH is used for this program.

    Assemble this program using ASM21.EXE and link using LD21.EXE to produce
    CONVOLV.EXE and CONVOLV.SYM. Load these files in the Simulator and execute.
    Check output sequence y(n) using data memory window.

}
.MODULE/RAM/ABS=0 CONVOLV;          {Beginning of CONVOLV Program}
.CONST M = 8;                       {Size of the h(n) sequence: set to 8}
.CONST N = 10;                      {Size of the x(n) sequence: set to 10}
.VAR/DM/CIRC dm_h[M];               {Sequence h(n) in DM as cir_buf}
.VAR/DM/CIRC dm_x[M+N-1];           {Sequence [0,...,0,x(n)] in DM as cir_buf}
.VAR/DM      dm_y[M+N -1];          {Output y in DM as linear_buf}
.INIT dm_h: <H(n).DAT>;             {Initialize h and x using data from}
.INIT dm_x: <X(n).DAT>;             { disk files}

        JUMP start; NOP; NOP; NOP; {Restart Interrupt}
        RTI; NOP;  NOP; NOP;            {Disable ALL other interrupts }
        RTI; NOP;  NOP; NOP;
        RTI; NOP;  NOP; NOP;
        RTI; NOP;  NOP; NOP;
        RTI; NOP;  NOP; NOP;
        RTI; NOP;  NOP; NOP;

start:  I0 = ^dm_h + M - 1;         {Index to Circular Buffer h @ M-1}
        L0 = %dm_h;                 {Length of Circular Buffer h }
        I1 = ^dm_x;                 {Index to Circular Buffer x }
        L1 = %dm_x;                 {Length of Circular Buffer x }
        I2 = ^dm_y;                 {Index to Circular Buffer y }
        L2 = 0;                     {Linear Buffer y }
        M0 = -1;                    {Post-decrement by 1 }
        M1 = N;                     {Post-increment by N }
        M2 = 1;                     {Post-increment by 1 }

        CNTR = M+N - 1;             {Set up counter for LOOP1 }
        DO loop1 until CE;          {Beginning of LOOP1 }

        CNTR = M;                   {Set up counter for LOOP2 }
        MR = 0;                     {Clear MR register }
        DO loop2 until CE;          {Beginning of LOOP2 }

        MX0 = DM(I0, M0 );           {Load MX0 register with input }
        MY0 = DM(I1, M2 );           {Load MY0 register with input }
loop2:  MR = MR + MX0 * MY0 ( SS );    {Do convolution }

        DM(I2,M2) = MR1;            {Store output }
loop1:  MODIFY( I1, M1 );           {Next data }

.ENDMOD;                            {End of CONVOLV Program }
```

———————— Listing 5-5: Convolution Program ————————

The program as usual begins with assembler directives. The .CONST directives assign values to the lengths of the respective sequences. The .VAR directives assign data memory buffers to the $h(n)$, $x(n)$, and $y(n)$ sequences. Even though this program performs linear convolution, *dm_h* and *dm_x* buffers are declared as circular buffers. It takes care of the addressing problem at the start of the convolution loop.

The new directive we have is .INIT, which initializes the data buffers from values stored in disk files. These files must be created by the user and should be located in the directory from which the Linker is invoked. The standard format for the data file is a single four- or six-character hexadecimal number per line of input (carriage returns are ignored). Refer to the *Cross-Software Manual* [2], appendix B for more details on the file format.

At this juncture, a discussion on the format of these two buffers, *dm_h* and *dm_x*, is in order. We are trying to simulate linear convolution using circular buffers. Referring to the above convolution equation, if we fold $h(m)$ to obtain $h(-m)$ and shift it with respect to $x(m)$, then we can obtain the result by the multiply/add operation over the overlap of $x(m)$ and $h(n-m)$. For the first $M-1$ sample points, $h(n-m)$ will partially overlap with $x(m)$. Hence some of its samples will multiply with zeros. This is what we have to simulate. This can be done by padding *dm_x* by $M-1$ zeros first followed by N values of $x(m)$. Therefore the length of *dm_x* buffer is $(N+M-1)$. Similarly, for the last $(M-1)$ sample points, $h(n-m)$ again partially overlaps with $x(m)$. This, however is correctly simulated by the circular nature of the *dm_x* buffer (why?).

Since this program has no external I/O, all interrupts except the RESTART interrupt are disabled. Following the *start:* label, initialization for all buffer address registers is done. Since we want to simulate $h(-m)$, I0 points to the bottom of the *dm_h* register and its index register is decremented by 1 using M0. L0 sets the length of the *dm_h* circular buffer. The index register I1 points to the top of the *dm_x* register and it is incremented by the M2 register during the convolution sum (in loop2), while I1 is incremented by N using M1 at the end of this sum (in loop1). This simulates the shifting of $h(n-m)$ by one sample (why?). The index register I2 points to the top of the *dm_y* buffer and is incremented by 1 using the M2 register. The L2 register initializes the *dm_y* buffer as a linear buffer.

The process of convolution now takes place inside two DO Loops. The outer *loop1:* computes $N+M-1$ samples of the convolution while the inner *loop2:* computes the convolution sum. Since the counter stack is four deep, the same CNTR can be used to regulate these DO Loops. The MR register is initialized before the accumulation of the sum. The $h(n)$ sequence values are brought in the MX0 register while the $x(n)$ sequence values are brought in the MY0 register. The two are multiplied and the result is accumulated in the MR register. Note that the default mode of operation for the MAC is the fractional 1.15 format arithmetic of 2's complement numbers. This is what is assumed in this program. At the end of loop2, the most significant fractional bits in MR1 are transferred to *dm_y* buffer and the process continues until all samples are processed.

We are now ready to simulate convolution in the Simulator using this program. Copy CONVOLV.DSP from the diskette to your working directory. Use EZLAB1.ACH architecture description file to assemble and link this program and generate CONVOLV.EXE and CON-VOLV.SYM files for the Simulator. Finally, create two data files; H(N).DAT containing 8 hexadecimal values, and X(N).DAT containing 17 hexadecimal values (with seven leading zeros as discussed above). Do not forget to follow the file format discussed in appendix B of the *Cross Software Manual* [2].

Invoke the Simulator and load the CONVOLV program. We can either single step through the Simulator to observe various register operations or execute the entire program. The program will stop when it tries to execute an undefined program memory location. At this time check the data memory corresponding to the *dm_y* buffer and verify that it does contain the proper convolution results. We have not provided any overflow protection in this program. Therefore depending on the values you choose in the data files (which should be fractions) we may get incorrect results.

EXPERIMENT 3A

In this experiment, we will develop a more detailed version of the above program to process speech signals in real-time.

1. Using the CONVOLV program as a guide, write EXPMT-3A.DSP program to incorporate the following features:

   - Input from the microphone and output to the speaker,
   - FIR filter coefficients in the Program Memory,
   - Input samples in the Data Memory,
   - Multifunction instruction inside the convolution loop,
   - Overflow protection (detection and saturation)

2. Using EXPMT-3A program, implement the following MA equation:

$$y(n) \; = \; \frac{1}{2} \sum_{m=0}^{4} \left(\frac{1}{2}\right)^m x(n-m)$$

Determine and plot the magnitude response of the above filter and verify it experimentally using a sinusoidal source.

3. Using the above filter, experiment with speech signals and record your observations.

5.4.2 Recursive Filter Program

An AR filter is a recursive (all-pole) filter described by the equation

$$y(n) \; = \; b_0 x(n) + \sum_{k=1}^{N} a_k y(n-k)$$

In this program using the Simulator, we will compute the impulse response of a first-order recursive filter given by

$$y(n) \; = \; x(n) + 0.5 y(n-1)$$

When $x(n)$ is an impulse sequence $\delta(n)$, the output of the filter $y(n)$ is equal to the impulse response $h(n)$. This is an infinite duration sequence, therefore the program computes only the first few samples. The appropriate program REC_FILT.DSP is shown in Listing 5-6.

```
{ ADSP-2101 Recursive Filter Program                              REC_FILT.DSP

  This program implements the following first-order recursive filter:
                      y(n) = x(n) + 0.5 * y(n-1)
  where y(n) is the output and x(n) is the input.  The input sequence is
  an impulse "delta(n)".  Hence the output y(n) is the impulse response
  of this filter.  The first 10 impulse response values are computed and
  stored in some data memory locations.  This program is intended for the
  Simulator. Architecture file EZLAB1.SYS is used for this program.

  Assemble using ASM21.EXE and link using LD21.EXE to produce REC_FILT.EXE
  and REC_FILT.SYM.  Load these files in the Simulator and execute.  Check
  the impulse response using data memory window.

}
.MODULE/RAM/ABS=0 REC_FILT;               {Beginning of REC_FILT Program}
.CONST samples=10;                        {Number of samples computed (10)}
.VAR/DM/CIRC dm_y[samples];               {Output y(n) buffer}

        JUMP start; NOP; NOP; NOP;        {Restart Interrupt}
        RTI; NOP; NOP; NOP;               {Disregard ALL other interrupts}
        RTI; NOP; NOP; NOP;
        RTI; NOP; NOP; NOP;
        RTI; NOP; NOP; NOP;
        RTI; NOP; NOP; NOP;
        RTI; NOP; NOP; NOP;

start:  I0 = ^dm_y;                       {I0 points to top of y buffer}
        L0 = %dm_y;                       {L0 = Length of buffer y}
        M0 = 1;                           {Post-increment by 1}

        CNTR = samples;                   {Set Counter for LOOP1}
        DO loop1 until CE;                {Initialize y buffer to 0}
loop1:  DM(I0, M0) = 0;

        MY0=0x4000;                       {Coefficient 0.5 in MY0}
        MR1=0x7FFF;                       {Set y(0) = 1 ~= 0.111 ... 1}

        CNTR = samples;                   {Set up counter for LOOP2}
        DO loop2 until CE;                        {Calculate difference eqn}
loop2:  DM(I0, M0) = MR1, MR=MR1 * MY0 ( SS ); {and store in y buffer}

.ENDMOD;                                  {End of REC_FILT Program}
```

———————————— Listing 5-6: Recursive Filter Program ————————————

This program is similar to the previous CONVOLV program in that it has no external input/output operations. Since $x(n) = \delta(n)$, we don't need a circular buffer for the input sequence. In fact, $\delta(n)$ can be simulated by choosing $x(n) = 0$ and $y(-1) = 1$ and then driving the recursive equation using this initial condition. Therefore the only circular buffer we need in this program is for the output sequence $y(n)$. The coefficient 0.5 is stored in MY0 while the MR register is initialized to 1. Note that in the fractional 1.15 format, the value 1 is approximated by 0x7FFF. The difference equation is simulated using a multifunction instruction and the DO Loop computes 10 samples of the impulse response.

Obtain a copy of REC_FILT.DSP from the diskette and load it in your working directory. Use the EZLAB1.ACH architecture description file to assemble and link this program, and generate REC_FILT.EXE and REC_FILT.SYM files for the Simulator. Invoke the Simulator and load the REC_FILT program. We can either single step through the Simulator to observe various register operations or execute the entire program. The program will stop when it tries to execute an undefined program memory location. At this time check the data memory corresponding to *dm_y* buffer and verify that it does contain proper impulse response results.

EXPERIMENT 3B

In this experiment, we will develop a more general (N^{th} order) AR filter program to process speech signals in real-time.

1. Using the REC_FILT program as a guide, write EXPMT-3B.DSP program to incorporate the following features:

   - Input from the microphone and output to the speaker,
   - AR filter coefficients (a_k's) in the Program Memory,
   - Output samples in the Data Memory,
   - Multifunction instruction inside the recursion loop,
   - Overflow protection (detection and saturation)

2. Using EXPMT-3B program, implement the following AR filter equation:

$$H(z) \;=\; \frac{K}{1+0.9z^{-1}+0.81z^{-2}}$$

   Determine and plot the pole-zero diagram and the magnitude response of the above filter. Determine the numerator K so that the maximum gain of the filter is equal to 1. Verify the operation of this filter experimentally using a sinusoidal source.

3. Using the above filter, experiment with speech signals and record your observations.

5.4.3 Exercises

1. Write a program to implement a 7-tap FIR filter with rectangular tap weights of equal heights ($b_n = \frac{1}{7}, 0 \le n \le 6$). Determine and plot the frequency response of this rectangular window and verify it using a sinusoidal source.

2. Consider the above 7-tap FIR filter in which the rectangular tap weights are replaced by triangualr tap weights given by:

$$b_n \;=\; \begin{cases} \dfrac{n}{7} & , \quad 0 \le n \le 3 \\[2ex] \dfrac{6-n}{7} & , \quad 4 \le n \le 6 \end{cases}$$

   Determine and plot the frequency response of this triangular window and verify it using a sinusoidal source.

3. Combine the programs EXPMT-3A.DSP and EXPMT-3B.DSP to implement a general Auto-Regressive Moving Average (ARMA) filter given by

$$y(n) \;=\; \sum_{m=0}^{M} b_m x(n-m) + \sum_{k=0}^{N} a_k y(n-k)$$

Using this program, implement the following All-Pass filter and verify its operation.

$$H(z) = \frac{0.72 - 1.7z^{-1} + z^{-2}}{1 - 1.7z^{-1} + 0.72z^{-2}}$$

4.   An interesting example of a stable recursive filter is reverberation in a good concert hall, whose impulse response may last for several seconds. This impulse response, $h(t)$, is given by

$$h(t) \quad = \quad \delta(t - \tau) + \alpha\delta(t - 2\tau) + \alpha^2\delta(t - 3\tau) + \ldots$$

where the concert hall feeds back τ seconds later a fraction, α, of its delayed input. For $\tau = 1$ ms and $\alpha = 0.993$, the reverberation time (when the response decays down to 60 dB) of 1 second can be achieved. This reverberation effect can be simulated on the ADSP-2101 system. The difference equation for the above impulse response is

$$y(n) = x(n - N) + \alpha y(n - N)$$

in which the delay in samples, N, depends on the sampling frequency F_s and is given by,

$$N \quad = \quad \tau F_s$$

Write a program to simulate this reverberation effect for an 8 KHz sampling frequency and demonstrate its results.

5.5 WAVEFORM GENERATION

Until now we studied operations that are necessary to implement the most basic DSP algorithms such as filtering. The common thread among all these algorithms is *shift*, *multiply* and *add*. However there are operations other than the above which are required in advance DSP applications. Consider an example of an adaptive filter used in echo cancellation. An echo canceller is implemented as an FIR filter whose filter coefficients are not constant but must be changed adaptively to track varying characteristics of a typical telephone channel. The coefficients are changed according to some optimum strategy which may require operations such as a square-root or a transcendental function. Sometimes we may have a need to generate pseudo random numbers to implement a random source. Therefore we need a procedure to implement these operations using basic computational blocks of the ADSP-2101 microcomputer.

Transcendental functions such as sines and logarithms are often approximated by polynomial expansions. The most widely used of these are the Taylor and Maclaurin series. They can be used to approximate almost any function whose derivative is defined over the specified input range. In this experiment we will develop subroutines to implement sine function approximation from a polynomial expansion and random number generation using the linear congruence method. Other transcendental functions can similarly be implemented using their appropriate polynomial expansions. Because the ADSP-2101 performs single precision (16-bit) fixed point operations, the accuracy of a polynomial expansion decreases as the order of the polynomial increases. Therefore the order of the polynomial must be limited to the minimum using optimized coefficients for the polynomials in the function approximation.

There are two important programming concepts in this experiment. First, the use of subroutines as a programming tool. We will implement these functions as subroutines and then call them up whenever needed. Therefore we will study how to interface the subroutines with a main program. To devise a waveform generator using functions, we need to generate function values at proper periodic time intervals, i.e., we need timing information. Hence the second important concept we will study is the use of the programmable interval Timer of the ADSP-2101 to generate periodic interrupts. Using these concepts we will generate and verify a single frequency tone at some frequency in Experiment 4A while in 4B we will produce and test a uniformly distributed random noise sequence.

5.5.1 Digital Sinewave Generator

The following formula approximates the sine of the input variable x:

$$\sin(x) \quad = \quad 3.140625x + 0.02026367x^2 - 5.325196x^3 + 0.5446778x^4 + 1.800293x^5$$

The approximation is accurate for any value of x from 0° to 90° (the first quadrant). However because $\sin(-x) = -\sin(x)$ and $\sin(x) = \sin(180° - x)$, the sine of any angle can be obtained from the sine of an angle in the first quadrant.

The subroutine, SINE.DSP, that implements this sine approximation, accurate to within two least significant bits (LSBs), is shown in Listing 5-7. This routine accepts input values in 1.15 format. The coefficients, which are initialized in data memory in 4.12 format, have been adjusted to reflect an input value scaled to the maximum range allowed by this format. On this scale, 180° equals the maximum positive value, 0x7FFF, and -180° equals the maximum negative value, 0x8000, as shown in Figure 5-1.

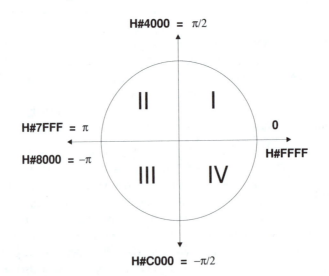

Figure 5-1: Scaled Angle Values

The routine shown in Listing 5-7 contains a new assembler directive called .ENTRY. It makes the label *sin:* visible to other programs (especially the main program) for use in subroutine calls or inter-module jumps. The routine then reads the scaled input angle from AX0. This angle is first modified to generate the angle in the first quadrant that will yield the same sine (or negative sine). If the input is in the second or fourth quadrants (bit 14 of the input value is a one) the input is negated to produce the twos complement, which represents an angle in the third or first quadrant, respectively. The sign bit of this angle is cleared to produce an angle in the first quadrant, and this result is stored in AR.

```
{ ADSP-2101 Sine Program                                          SINE.DSP

    This subroutine calculates sine value, Y = sin( X ), by using a polyno-
    mial approximation:
    sin(x) = 3.140625x+0.02026367x^2-5.325196x^3+0.5446778x^4+1.800293x^5

    Calling Parameters
         AX0 = X in scaled 1.15 format
         M3 = 1
         L3 = 0
    Return Value
         AR = Y in 1.15 format

}

.MODULE/RAM SINE_FUNCTION;              { Beginning of SINE Program }
.VAR/DM sin_coeff[5];                  { Sine Polynomial Coefficients }
.INIT sin_coeff : H#3240, H#0053, H#AACC, H#08B7, H#1CCE;
                                       { Initialize coefficients }
.ENTRY  sin;                           { Entry point of sine function }

sin:     I3=^sin_coeff;                { Index to coefficients }
         AY0=H#4000;
         AR=AX0, AF=AX0 AND AY0;       { Check 2nd or 4th quad. }
         IF NE AR='-AX0;               { If yes, negate input }
         AY0 = H#7FFF;
         AR = AR and AY0;              { Remove sign bit }
         MY1=AR;
         MF=AR*MY1 ( RND ), MX1=DM( I3, M3 ); { MF= X * X }
         MR=MX1*MY1 ( SS ), MX1=DM( I3, M3 ); { MR= C1 * X }
         CNTR=3;
         DO approx UNTIL CE;           { Calculate sine value }
           MR=MR+MX1*MF (SS);
approx:    MF=AR*MF (RND ), MX1=DM( I3, M3 );
         MR=MR+MX1*MF (SS);
         SR=ASHIFT MR1 BY 3 ( HI );    { Convert to 1.15 format }
         SR = SR OR LSHIFT MR0 BY 3 ( LO );
         AR=PASS SR1;
         IF LT AR=PASS AY0;            { Saturate if needed }
         AF=PASS AX0;
         IF LT AR=-AR;                 { Negate if needed }
         RTS;
.ENDMOD;                               { End of SINE Program }
```

—————————— Listing 5-7: Sine Approximation ——————————————

If the original angle is in the first quadrant, its value is unchanged. If it is in the second quadrant, negation changes it to the third quadrant, and the sign bit removal changes it to the first quadrant. If the original angle is in the third quadrant, the removal of the sign bit changes it to the first quadrant. An angle that is originally in the fourth quadrant is changed to the first quadrant by negation.

The sine of the modified angle is calculated by multiplying increasing powers of the angle by the appropriate coefficients. The square of the angle is computed and stored in MF while the first coefficient is fetched from data memory. The first term of the sine approximation is stored in the MR registers (in which the result is subsequently accumulated) in parallel with the second coefficient fetch. In the *approx:* loop, the next term of the approximation is computed and added to the partial result in MR; then the multiplication instruction fetches the next coefficient and generates the next power of the angle at the same time.

Because the coefficients are in 4.12 format, a shift instruction is needed to scale the result to 1.15 format. The result is then checked for overflow. If the value in SR1 exceeds 0x7FFF, the routine saturates the result at the maximum positive value, 0x7FFF, which is read from AY0. Then the sign of the result is restored, if necessary. If the input angle (stored in AX0) is negative, the result must be negated.

This routine requires 25 cycles to generate one sine value. At 12.5 Mhz processor speed this means that one sine value is available in 2 microseconds.

A simple simulator main program to generate $\sin(\pi/2)$ using the above *sine* subroutine is shown in Listing 5-8. It contains the assembler directive .EXTERNAL to interface a subroutine with the main program. This directive assigns the external attribute to the *sin:* label declared in another module using the .ENTRY directive. Therefore it is now possible to call the sine subroutine from the main program. Since the calling parameters of this subroutine are AX0, M3, and L3 these must be initialized before the subroutine is called. AX0 is set to 0x4000 which corresponds to $\pi/2$ radians. After the program returns from the subroutine, the sine value will be in the AR register in 1.15 format.

```
.MODULE/RAM/BOOT=0/ABS=0 MAIN;
.EXTERNAL sin;

            JUMP start; NOP; NOP; NOP;
            RTI; NOP; NOP; NOP;
            RTI; NOP; NOP; NOP;
            RTI; NOP; NOP; NOP;
            RTI; NOP; NOP; NOP;
            RTI; NOP; NOP; NOP;
            RTI; NOP; NOP; NOP;

start:      AX0=0x4000;
            M3=1;
            L3=0;
            call sin;

.ENDMOD;
```

──────────── **Listing 5-8: Main Program** ────────────

To generate one executable file, assemble the main program and the subroutine separately. Then link the two using the EZLAB1.ACH architecture description file. Finally execute the main program in the simulator. Verify that the correct sine of $\pi/2$ is in the register AR1.

5.5.2 TIMER Operation

So far we generated one sine value using the subroutine. To produce a sine waveform we have to compute several sine values over the 0 to 2π interval (i.e. over one cycle) and generate these samples at periodic intervals. The time interval between two consecutive samples determines the frequency of the waveform. This periodic time interval generation is accomplished by the Timer unit of the processor. Therefore we have to study the Timer in detail in order to understand its operation.

The ADSP-2101 Timer includes two 16-bit registers, TCOUNT and TPERIOD and one 8-bit register, TSCALE. These registers are memory mapped: TPERIOD at 0x3FFD, TCOUNT at 0x3FFC and TSCALE at 0x3FFB. The mode control instruction enables and disables the timer by setting and clearing bit 5 in the mode status register, MSTAT. The Timer must be enabled before its counting capabilities can be used.

TPERIOD is the period register which holds the period of the interrupt in cycles. TCOUNT is the count register. When the timer is enabled, TCOUNT is decremented as often as once every instruction cycle. When the counter reaches zero, an interrupt is generated. TCOUNT is then reloaded from the TPERIOD register and the count begins again. TSCALE stores a scaling value that is one less than the number of cycles between decrements of TCOUNT. For example, if the value in TSCALE register is 0, the counter register decrements once every cycle. If the value in TSCALE is 1, the counter decrements once every two cycles. Therefore using these three registers, interrupts from 5.24 ms (when TPERIOD is at maximum and TSCALE at minimum with resolution of 80 ns) up to 1.34 seconds (when both TPERIOD and TSCALE are maximum with resolution of 20.48μs) at 80ns cycle time can be generated.

EXPERIMENT 4A

In this experiment we will generate a sinusoidal waveform at 3 KHz frequency with 16 samples per cycle and display it on the scope.

1. Using the SINE.DSP subroutine, write EXPMT-4A.DSP program to incorporate the following features:
    *   Output to the DAC0 port,
    *   16 samples per cycle. Store 16 angle values in Data memory as a circular buffer and initialize them from disk file.
    *   Properly set Timer registers for 16 samples per cycle and 3 KHz frequency.
    *   Enable Timer interrupt by setting bit 0 in IMASK.
    *   Enable Timer by setting bit 5 in the mode status register MSTAT.

2. Using EXPMT-4A program, generate the sinusoidal waveform, display it on the scope and verify that the generated frequency is correct.

3. Also verify the generated frequency by "listening" to it on the speaker and comparing it with one generated using the function generator.

5.5.3 Uniform Random Number Generator

Although the generation of a random number is not, strictly speaking, a function, it is a useful operation for many applications. One such application is in high-speed modems, in which it can be used as a training signal for the adaptive equalizer. The means for generating random numbers on a digital computer, of course, is by the computation of a function that approximates the random number. Many such functions have been proposed. The implementation presented here is based on the linear congruence method, which uses the following equation.

$$x(n+1) \quad = \quad (ax(n)+c) \bmod m$$

The initial value of x, $x(0)$, is called the seed value and is generally not important, because with a good choice of a and c all m values are generated before the output sequence repeats. The random number sequence produced by the above equation is thus uniform in the sense that the output is uniformly distributed between 0 and $m - 1$. Of course different seed values should be used at different times if different sequences are desired. By choosing the modulus $m = 2^{32}$, one can ensure a long sequence and have a convenient modulus for the ADSP-2101.

Listing 5-9 shows the ADSP-2101 routine used to compute random numbers based on the linear congruence method. The values of a and c that are used in this program ($a = 1664525$) and $c = 32767$ were chosen according to the rules given in Knuth [7]. The initial seed value is stored in the SR register before the subroutine is called for the first time and the first number produced by this routine is the initial seed value. The random numbers are returned in the AR register while the subsequent seed values are in the SR register. Note that, although only the most significant 16 bits of the 32-bit x value are used as random numbers in this routine, any or all of the bits can be used. However when using a value of m equal to the word size of the machine, the least significant bits of $x(n)$ are much less random than the most significant bits. Thus, one should always use the b most significant bits when only a b-bit random number is desired.

```
{ ADSP-2101 Random Subprogram                                RANDOM.DSP

     This subroutine calculates uniformly distributed random numbers using a
linear congruence method

     Calling Parameters
          SR1 = MSW of the seed value
          SR0 = LSW of the seed value

     Return Values
           AR = The Random Number
          SR1 = MSW of the new seed value
          SR0 = LSW of the new seed value

     Altered Registers
          MY0, MY1, MR, SI, SR, AR

     Computation Time
          10*N + 4 cycles

     Note: The first time the routine is called it will return the portion
           of the seed that is in SR1.  Subsequent times it will return a
           pseuo-random number in AR.
```

```
}

.MODULE/RAM URAND_SUB;                { Beginning of RANDOM Subprogram }
.ENTRY   urand;                       { Entry point of random function }

urand:   MY1=25;                      { Upper half of a }
         MY0=26125;                   { Lower half of a }
         AR=SR1, MR=SR0*MY1(UU);      { a(hi) * x(lo) }
         MR=MR+SR1*MY0(UU);           { a(hi) * x(lo) + a(lo) * x(hi) }
         SI=MR1;
         MR1=MR0;
         MR2=SI;
         MR0=H#FFFE;                  { c-32767, left-shifted by 1 }
         MR=MR+SR0*MY0(UU);           { ( above ) + a(lo) X x(lo) + c }
         SR=ASHIFT MR2 BY 15 ( HI );
         SR=SR OR LSHIFT MR1 BY -1(HI);{ right shift by 1 }
         SR=SR OR LSHIFT MR0 BY -1(LO);
         RTS;

.ENDMOD;                              { End of RANDOM Program }
```

──────────── **Listing 5-9: Random Number Generator** ────────────────────

The routine requires $10N + 4$ cycles to execute, where N is the number of random numbers desired. For example, computing 2^{16} (65,536) random numbers using a 12.5 MHz ADSP-2101 takes 52.43 milliseconds. Computing all $m=2^{32}$ numbers in the sequence requires almost one hour.

EXPERIMENT 4B

In this experiment we will generate a random sequence at 8KHz sampling rate.

1. Using RANDOM.DSP subroutine write EXPMT-4B.DSP program to generate random sequence. Adjust the timer registers to generate a sequence at an 8 Khz sampling rate. Output this random sequence to the SPORT0 transmit register.

2. Connect the speaker to the SPORT0 transmit register and listen to the output. Comment on the auditory qualities of the noise.

5.5.4 Exercises

1. The sinusoidal waveform can also be generated using a second order recursive filter given by

$$y(n) \quad = \quad x(n) + 2\cos\omega_0 y(n-1) - y(n-2)$$

When this filter is excited by an impulse, i.e. $x(n) = \delta(n)$, it generates a digital sine wave with digital frequency of ω_0 rad/sam since it has two poles on the unit circle at conjugate angles ω_0. The frequency of the analog sine wave generated by this filter is given by

$$F \quad = \quad \frac{F_s\omega_0}{2\pi}$$

where F_s is the sampling frequency. Using the programs developed in Experiment 3, write a program to generate a 1 KHz tone. Note that one of the filter coefficient can range between -2 and 2. Hence implement a 2.14 fixed-point format representation in your program. Comment on any possible problems with this approach in waveform generation.

2. Write a program to generate a noisy sinusoidal waveform by incorporating both the sine and the random number generators. Generate a 2 KHz waveform corrupted by noise distributed uniformly between -1 and 1. Implement a 2.14 fixed-point format in your program.

3. To generate random numbers from a Gaussian distribution one can add 12 numbers from the [0, 1] uniform generator and subtract 6. The mean of the uniform generators is 1/2 so that the subtraction of 6 from the sum gives a mean of exactly 0. Since the variance of the original uniform distribution is 1/12 the 12 independent numbers give a distribution with variance exactly 1. Write a program to generate Gaussian random numbers with mean 0 and variance 1.

5.6 SUMMARY

In this chapter, we described several simple experiments as a vehicle to effectively program and use the microcomputer. In the process we also implemented different DSP operations and analyzed the results in terms of the expected and achieved outcomes. What we studied is the essence of DSP because more complicated operations can be achieved using these simple programs. We can now treat this ADSP-2101 microcomputer as a tool to generate more sophisticated algorithms and develop useful and interesting application. This is what we will do in the next five chapters.

chapter 6

FIR FILTER IMPLEMENTATIONS

6.1 INTRODUCTION

Digital signal processing comprises of two important areas: signal filtering in the time domain and signal representation in the frequency domain. With the programming background that we have developed in the last chapter, we are now in a position to treat the subject of digital filter design and implementation on the ADSP-2101. As described in Chapter 5, there are two types of digital filters namely FIR and IIR. This chapter deals with the design and implementation of FIR filters. A similar treatment for IIR filters is given in the next chapter. In Chapter 8, we describe signal representation in the form of the Discrete Fourier Transform (DFT) and an efficient method for computing the DFT, called the Fast Fourier Transform Algorithm.

In practice, FIR filters are employed in filtering problems where there is a requirement for a linear phase characteristic within the passband of the filters. Using linear phase FIR filters it is possible to implement such systems as differentiators and Hilbert transformers, which can not be implemented using IIR filters. However, if phase distortion is unimportant then IIR filters are generally preferable. In this chapter, we consider FIR filters which are of a frequency-selective type, i.e., we will design and implement filters which are multiband filters with magnitude response specified in each band. There is another class of filters in which frequency-domain characteristics of the desired filters are not specified explicitly. These filters include Wiener filters, inverse filters for deconvolution, and equalizers, in which the design criterion is specified in terms of minimizing some performance measure in the time-domain. Some of these filters will be considered later in Chapter 10. From an implementation point of view, however, this classification is unimportant.

We begin in Section 6.2 with a brief discussion on frequency-selective linear phase FIR filters. We describe window design technique and the Parks-McClellan algorithm as two representative methods of FIR filter design. In Section 6.3, we describe the implementation of FIR filters using single-precision arithmetic. In some applications, the 16-bit precision of the ADSP-2101 microcomputer may not be enough. Therefore, we describe a double-precision implementation in Section 6.4. Section 6.5 deals with the implementation of all-zero lattice filters. Finally, in Section 6.6 we present a single sideband modulator as an example of a complete system implementation involving many components besides an FIR filter.

6.2 OVERVIEW OF FIR FILTER DESIGN

FIR filter has several design advantages over an Infinite-duration Impulse Response (IIR) digital filter:
- it is always stable,
- it is always realizable, and
- it can always be designed to have exact linear phase.

The third advantage makes the FIR filter an indispensable choice in applications requiring no delay distortion but only fixed delay. Additionally, the design problem for linear phase FIR filters requires only real arithmetic which is easy to implement. Therefore in this overview, we discuss design techniques and examples of linear phase FIR filters.

As discussed in Chapter 4, an FIR filter is described by the difference equation (or convolution formula)

$$y(n) \quad = \quad \sum_{k=0}^{N-1} h(k)x(n-k) \tag{6-1}$$

where $h(n), n = 0, \ldots, N-1$ is the impulse response of the filter. The frequency response of the FIR filter is given by

$$H(e^{j\omega}) \quad = \quad \sum_{n=0}^{N-1} h(n)e^{-j\omega n}$$

The magnitude of the frequency response, $|H(e^{j\omega})|$, is called the magnitude response while the angle, $\angle H(e^{j\omega})$, is called the phase response. If we impose a linear phase constraint on the phase response in the form

$$\angle H(e^{j\omega}) \quad = \quad \beta - \alpha\omega, \quad \omega \geq 0$$

then it can be shown [7] that we have the following two solutions:
- either

$$h(n) = h(N-1-n), \quad 0 \leq n \leq N-1,$$

$$\beta = 0, \quad \text{and} \quad \alpha = \frac{N-1}{2}$$

- or

$$h(n) = -h(N-1-n), \quad 0 \leq n \leq N-1,$$

$$\beta = \pm\frac{\pi}{2}, \quad \text{and} \quad \alpha = \frac{N-1}{2}$$

The first solution results in the impulse response $h(n)$ being symmetric about α. All multiband (lowpass, highpass, bandpass, or bandstop) linear-phase FIR filters exhibit this behavior. The second solution implies that the impulse response $h(n)$ is antisymmetric about α, a behavior exhibited by FIR differentiators and Hilbert transformers. These symmetry conditions can be used to reduce multiplications by about 50% in implementing FIR filters. However in our ADSP-2101 implementations, multiplication and addition can be done in one instruction cycle. Therefore symmetry conditions are of little importance from implementation point of view.

The linear-phase FIR filters are specified in terms of their magnitude response. The essence of FIR filter design is then to determine a causal impulse response $h(n)$ to approximate the given magnitude specifications subject to symmetry conditions on $h(n)$. The phase response then can be determined from the length and the symmetry of the impulse response. In the following sections we briefly describe two well known approaches to FIR filter design: window design method and Parks-McClellan algorithm.

6.2.1 Window Design Method

The simplest method of designing a linear phase FIR filter is the window design method. The given magnitude specifications can be thought of as an ideal filter $H_D(e^{j\omega})$ to be approximated. For example, an ideal lowpass filter has unity magnitude response and linear phase over the passband and zero response over the stopband:

$$H_D(e^{j\omega}) = \begin{cases} 1 \cdot e^{-j\alpha\omega} & , \quad |\omega| \le \omega_c \\ 0 & , \quad \omega_c < |\omega| \le \pi \end{cases} \tag{6-2}$$

where ω_c is called the corner frequency. Clearly the impulse response of the ideal filter, $h_d(n)$, is of infinite duration given by

$$h_d(n) = \frac{1}{2\pi} \int_{-\pi}^{\pi} H_d(e^{j\omega}) e^{j\omega n} d\omega \tag{6-3}$$

To obtain a causal FIR filter of length N, we may truncate $h_d(n)$, i.e.,

$$h(n) = \begin{cases} h_d(n) & , \quad 0 \le n \le N-1 \\ 0 & , \quad otherwise \end{cases}$$

and

$$\alpha = \frac{N-1}{2}$$

This operation is called *windowing*. In general, $h(n)$ can be thought of as being formed by the product of $h_d(n)$ and a *window function*, $w(n)$, as follows:

$$h(n) = h_d(n) \cdot w(n) \tag{6-4}$$

and

$$w(n) = \begin{cases} \text{some symmetric function over} & 0 \le n \le N-1 \\ 0 & , otherwise \end{cases}$$

Depending upon how $w(n)$ is specified, one obtains different window designs. Some commonly used window types and their functions are shown in Table 6-1. In Table 6-2, we provide a summary of window function characteristics in terms of transition width (as a function of N) and the minimum stopband attenuation (in DB). These tables along with the given specifications can be used to obtain the approximated impulse response $h(n)$. In the remainder of this section, we describe the window design technique by using examples of some representative digital filters.

Window Type	Window Function $w(n), 0 \leq n \leq N-1$
Rectangular	1
Bartlett	$\dfrac{2n}{N-1}, \quad 0 \leq n \leq \dfrac{N-1}{2}$ $2 - \dfrac{2n}{N-1}, \quad \dfrac{N-1}{2} \leq n \leq N-1$
Hanning	$0.5\left(1 - \cos\left(\dfrac{2\pi n}{N-1}\right)\right)$
Hamming	$0.54 - 0.46\cos\left(\dfrac{2\pi n}{N-1}\right)$
Blackman	$0.42 - 0.5\cos\left(\dfrac{2\pi n}{N-1}\right) + 0.08\cos\left(\dfrac{4\pi n}{N-1}\right)$
Kaiser	$\dfrac{I_0\left[\beta\sqrt{1 - \left(1 - \frac{2n}{N-1}\right)^2}\right]}{I_0[\beta]}$ where I_0 is the modified Bessel function of order zero.

Table 6-1: Window Functions ───────────

Window Type	Transition Width $\Delta\omega$	Minimum Sideband Attenuation in DB
Rectangular	$\dfrac{1.8\pi}{N}$	21
Bartlett	$\dfrac{5.6\pi}{N}$	25
Hanning	$\dfrac{6.2\pi}{N}$	44
Hamming	$\dfrac{6.6\pi}{N}$	53
Blackman	$\dfrac{11\pi}{N}$	74
Kaiser (β=4.54)	$\dfrac{5.8\pi}{N}$	50
Kaiser (β=5.67)	$\dfrac{7.8\pi}{N}$	60

Table 6-2: Window Function Characteristics ───────────

Example 6-1:

In the first example, we will design a lowpass filter for our EZ-LAB target system. The codec in the EZ-LAB target system operates at an 8 KHz sampling rate. Therefore we design our filter for the following specifications:

$$\begin{aligned}
\text{Sampling frequency:} &\quad 8\text{ KHz} \\
\text{Passband:} &\quad 0 - 1\text{ KHz} \\
\text{Stopband:} &\quad 1.4 - 4\text{ KHz} \\
\text{Stopband Attenuation:} &\quad 50\text{ DB}
\end{aligned}$$

From Table 6-2 we observe that both the Hamming and Blackman window functions can provide attenuation of more than 50 DB. Let us choose the Hamming window which is the better of the two. The required transition width is 0.4 KHz or

$$\Delta\omega \;=\; \frac{2\pi \cdot 400}{8000} \;=\; 0.1\pi$$

From Table 6-2, using the transition width column, we obtain the filter length N as

$$N \;=\; \frac{6.6\pi}{0.1\pi} \;=\; 66$$

We choose $N = 67$ which gives $\alpha = \frac{N-1}{2} = 33$. The corner frequency of the ideal lowpass filter is

$$\omega_c \;=\; \frac{\left(\frac{1000+1400}{2}\right)2\pi}{8000} \;=\; 0.3\pi$$

From equation (6-2), the ideal frequency response is

$$H_D(e^{j\omega}) = \begin{cases} 1 \cdot e^{-j33\omega} & ,\quad |\omega| \le 0.3\pi \\ 0 & ,\quad 0.3\pi < |\omega| \le \pi \end{cases}$$

and from equation (6-3), the ideal impulse response is

$$h_d(n) \;=\; \frac{\sin[0.3\pi(n-33)]}{\pi(n-33)}$$

From Table 6-1, the window function for Hamming window is

$$w(n) \;=\; 0.54 - 0.46\cos(2\pi n/66), \quad 0 \le n \le 66$$

Finally from equation (6-4), the impulse response of the designed FIR filter is

$$h(n) \;=\; \frac{\sin[0.3\pi(n-33)]}{\pi(n-33)} \cdot [0.54 - 0.46\cos(2\pi n/66)], 0 \le n \le 66$$

These FIR filter coefficients are shown in Table 6-3 and the frequency response is shown in Figure 6-1.

```
                              Filter Length = 67
                              ------------------
h(  0) = -2.384527E-04 = h(66)     h(17) =   6.570477E-03  = h(49)
h(  1) = -7.765305E-04 = h(65)     h(18) =   1.284836E-02  = h(48)
h(  2) = -7.336193E-04 = h(64)     h(19) =   8.665973E-03  = h(47)
h(  3) = -3.645996E-09 = h(63)     h(20) =  -5.224195E-03  = h(46)
h(  4) =  1.002993E-03 = h(62)     h(21) =  -1.844362E-02  = h(45)
h(  5) =  1.417815E-03 = h(61)     h(22) =  -1.802628E-02  = h(44)
h(  6) =  5.574838E-04 = h(60)     h(23) =  -3.472296E-08  = h(43)
h(  7) = -1.283883E-03 = h(59)     h(24) =   2.407036E-02  = h(42)
h(  8) = -2.636652E-03 = h(58)     h(25) =   3.303238E-02  = h(41)
h(  9) = -1.861351E-03 = h(57)     h(26) =   1.266898E-02  = h(40)
h( 10) =  1.168261E-03 = h(56)     h(27) =  -2.890585E-02  = h(39)
h( 11) =  4.265746E-03 = h(55)     h(28) =  -6.040657E-02  = h(38)
h( 12) =  4.278595E-03 = h(54)     h(29) =  -4.523324E-02  = h(37)
h( 13) =  1.676480E-08 = h(53)     h(30) =   3.217674E-02  = h(36)
h( 14) = -5.849049E-03 = h(52)     h(31) =   1.501070E-01  = h(35)
h( 15) = -7.980914E-03 = h(51)     h(32) =   2.569817E-01  = h(34)
h( 16) = -2.997856E-03 = h(50)     h(33) =   3.000000E-01  = h(33)
```

Table 6-3: Example 6-1 Filter Coefficients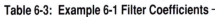

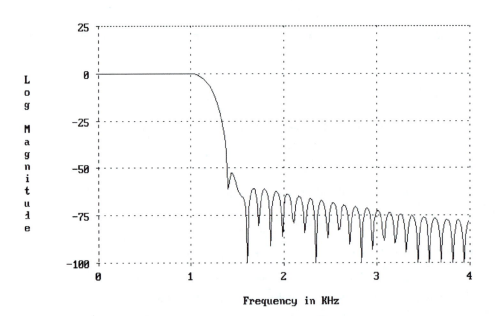

Figure 6-1: Example 6-1 Frequency Response

Example 6-2:

In this example, we will design a bandpass filter for the following specifications:

Sampling frequency:	8 KHz
Lower Stopband:	0 – 0.6 KHz
Passband:	1 – 2 KHz
Upper Stopband:	2.4 – 4 KHz
Stopband Attenuation:	40 DB
Window Function:	Hanning Window

Following the procedure in Example 6-1, we obtain the following results:

Transition width = Min[1-0.6, 2.4-2] = 0.4 KHz or $\Delta\omega = 0.1\pi$. Hence

$$N \;=\; \frac{6.2\pi}{0.1\pi} \;=\; 62.$$

Choose $N = 63$, then $\alpha = 31$.

Corner Frequencies: 0.8 KHz and 2.2 KHz or $\Delta\omega_{c1} = 0.2\pi$ and $\Delta\omega_{c1} = 0.55\pi$

Ideal frequency response:

$$H_D(e^{j\omega}) = \begin{cases} 1 \cdot e^{-j31\omega} & , \quad 0.2\pi \le |\,\omega\,| \le 0.55\pi \\ 0 & , \quad otherwise \end{cases}$$

Ideal impulse response:

$$h_d(n) \;=\; \frac{\sin[0.55\pi(n-31)]}{\pi(n-31)} - \frac{\sin[0.2\pi(n-31)]}{\pi(n-31)}$$

Hanning window function:

$$w(n) \;=\; 0.5 - 0.5\cos(2\pi n/62), \quad 0 \le n \le 62$$

Finally, the FIR filter coefficients are:

$$h(n) \;=\; \frac{\sin[0.55\pi(n-31)] - \sin[0.2\pi(n-31)]}{\pi(n-31)} \cdot \left[0.5 - 0.5\cos\left(\frac{2\pi n}{62}\right)\right], \quad 0 \le n \le 62$$

These FIR filter coefficients are shown in Table 6-4 and the frequency response is shown in Figure 6-2.

```
                          Filter Length = 63
    h( 0) = -2.218129E-16 = h(62)  |  h(16)  =   7.882669E-03 = h(46)
    h( 1) =  2.721886E-05 = h(61)  |  h(17)  =  -1.828366E-02 = h(45)
    h( 2) =  4.845893E-05 = h(60)  |  h(18)  =  -2.151320E-02 = h(44)
    h( 3) = -4.195137E-10 = h(59)  |  h(19)  =  -1.917156E-08 = h(43)
    h( 4) =  6.712061E-04 = h(58)  |  h(20)  =  -8.989551E-03 = h(42)
    h( 5) =  1.074378E-03 = h(57)  |  h(21)  =  -2.433423E-02 = h(41)
    h( 6) = -8.068400E-04 = h(56)  |  h(22)  =   2.121634E-02 = h(40)
    h( 7) = -1.880663E-03 = h(55)  |  h(23)  =   6.391277E-02 = h(39)
    h( 8) = -1.292477E-04 = h(54)  |  h(24)  =   1.987664E-02 = h(38)
    h( 9) = -1.801648E-03 = h(53)  |  h(25)  =  -1.068490E-02 = h(37)
    h(10) = -5.624267E-03 = h(52)  |  h(26)  =   4.218759E-02 = h(36)
    h(11) = -2.593494E-08 = h(51)  |  h(27)  =   7.281579E-08 = h(35)
    h(12) =  8.613655E-03 = h(50)  |  h(28)  =  -1.909672E-01 = h(34)
    h(13) =  4.253979E-03 = h(49)  |  h(29)  =  -1.984944E-01 = h(33)
    h(14) =  4.770481E-04 = h(48)  |  h(30)  =   1.269665E-01 = h(32)
    h(15) =  1.110132E-02 = h(47)  |  h(31)  =   3.500000E-01 = h(31)
```

Table 6-4: Example 6-2 Filter Coefficients ───────────────────────────────

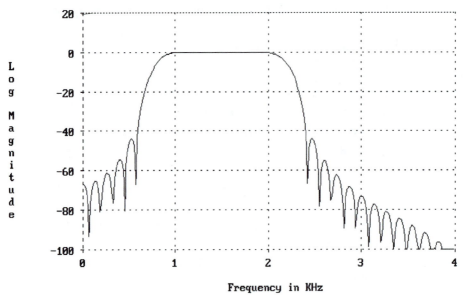

Figure 6-2: Example 6-2 Frequency Response

Example 6-3:

In this example, we will design a 40-tap FIR differentiator using a Blackman window. Differentiators have antisymmetric impulse responses and their length N must an even number so that their frequency response is not zero at $\omega = \pi$. The ideal frequency response of a typical differentiator is given by

$$H_D(e^{j\omega}) = \frac{j \, |\omega|}{\pi} e^{-j\alpha\omega}, -\pi < \omega \leq \pi$$

From equation (6-3), the ideal impulse response is

$$h_d(n) = \left\{ \begin{matrix} \dfrac{-1}{\pi^2(n-\alpha)^2} & (n-\alpha) \geq 0 \\[2ex] \dfrac{1}{\pi^2(n-\alpha)^2} & (n-\alpha) < 0 \end{matrix} \right\}$$

Note that since N is even, α and hence $(n-\alpha)$ are not integers. In this example, $\alpha = (40-1)/2 = 19.5$. Using the Blackman window function from Table 6-1, the FIR differentiator coefficients are given by

$$h(n) = h_d(n) \cdot \left[0.42 - 0.5\cos\left(\frac{2\pi n}{39}\right) + 0.08\cos\left(\frac{4\pi n}{39}\right) \right], 0 \leq n \leq 39$$

These coefficients are shown in Table 6-5 and the frequency response is shown in Figure 6-3. Note the distortion effect at the high frequency end, resulting from our specification that $|H_D(e^{j\omega})|$ be linear out to $\omega = \pi$.

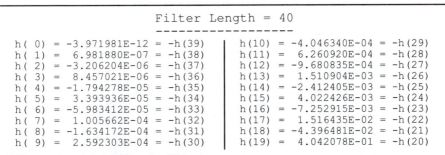

```
                        Filter Length = 40
                        ------------------
h( 0) = -3.971981E-12 = -h(39)    h(10) = -4.046340E-04 = -h(29)
h( 1) =  6.981880E-07 = -h(38)    h(11) =  6.260920E-04 = -h(28)
h( 2) = -3.206204E-06 = -h(37)    h(12) = -9.680835E-04 = -h(27)
h( 3) =  8.457021E-06 = -h(36)    h(13) =  1.510904E-03 = -h(26)
h( 4) = -1.794278E-05 = -h(35)    h(14) = -2.412405E-03 = -h(25)
h( 5) =  3.393936E-05 = -h(34)    h(15) =  4.022426E-03 = -h(24)
h( 6) = -5.983412E-05 = -h(33)    h(16) = -7.252915E-03 = -h(23)
h( 7) =  1.005662E-04 = -h(32)    h(17) =  1.516435E-02 = -h(22)
h( 8) = -1.634172E-04 = -h(31)    h(18) = -4.396481E-02 = -h(21)
h( 9) =  2.592303E-04 = -h(30)    h(19) =  4.042078E-01 = -h(20)
```

Table 6-5: Example 6-3 Filter Coefficients —————————————————————————

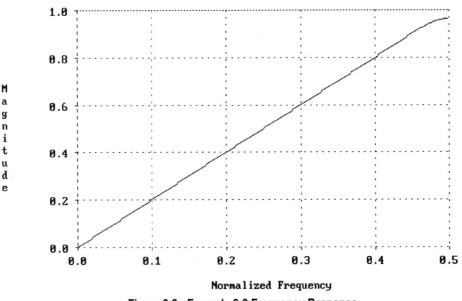

Figure 6-3: Example 6-3 Frequency Response

Example 6-4:

In this last example, we will design a Hilbert Transformer which also has an antisymmetric impulse response but the filter order N must be an odd number. We will use a Hamming window to design a 51-tap Hilbert transformer. The ideal frequency response of the Hilbert transformer is given by

$$H_D(e^{j\omega}) = \begin{cases} -j, & 0 < \omega \leq \pi \\ +j, & -\pi < \omega < 0 \end{cases}$$

From equation (6-3), the ideal impulse response is

$$h_d(n) = \begin{cases} \dfrac{2}{\pi(n-\alpha)} & (n-\alpha) \quad odd \\ 0 & (n-\alpha) \quad even \end{cases}$$

Since $N = 51$, $\alpha = 25$. Hence $(n - \alpha)$ is an integer. Using the Hamming window function, the FIR Hilbert transformer coefficients are given by

$$h(n) = \frac{2}{\pi(n-25)} \cdot \left[0.54 - 0.46 \cos\left(\frac{2\pi n}{50}\right) \right], n = 0, 2, ..., 50$$

These coefficients are shown in Table 6-6 and the frequency response is shown in Figure 6-4. Note that the Hilbert transformer has nulls in the frequency response characteristics at $\omega = 0$ and $\omega = \pi$.

```
                        Filter Length = 51
                        -------------------
   h( 0)  =  -2.037183E-03  =  h(50)      h(13)  =  -1.153552E-13  =  h(37)
   h( 1)  =  -3.391498E-14  =  h(49)      h(14)  =  -3.624076E-02  =  h(36)
   h( 2)  =  -2.614341E-03  =  h(48)      h(15)  =  -1.152686E-13  =  h(35)
   h( 3)  =  -2.248802E-14  =  h(47)      h(16)  =  -5.205134E-02  =  h(34)
   h( 4)  =  -4.150121E-03  =  h(46)      h(17)  =  -1.063189E-13  =  h(33)
   h( 5)  =  -5.672695E-14  =  h(45)      h(18)  =  -7.577731E-02  =  h(32)
   h( 6)  =  -6.857882E-03  =  h(44)      h(19)  =  -8.874694E-14  =  h(31)
   h( 7)  =  -1.318217E-13  =  h(43)      h(20)  =  -1.161383E-01  =  h(30)
   h( 8)  =  -1.099179E-02  =  h(42)      h(21)  =  -6.374568E-14  =  h(29)
   h( 9)  =  -9.304425E-14  =  h(41)      h(22)  =  -2.053517E-01  =  h(28)
   h(10)  =  -1.688537E-02  =  h(40)      h(23)  =  -3.330737E-14  =  h(27)
   h(11)  =  -3.634993E-14  =  h(39)      h(24)  =  -6.343105E-01  =  h(26)
   h(12)  =  -2.502975E-02  =  h(38)           h(25)  =    0  =  h(25)
```

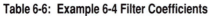

Table 6-6: Example 6-4 Filter Coefficients

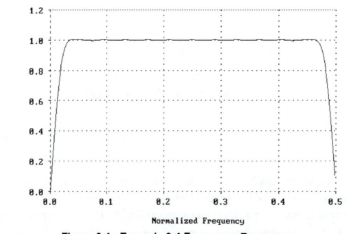

Normalized Frequency

Figure 6-4: Example 6-4 Frequency Response

6.2.2 Parks-McClellan Algorithm

The window design technique which we discussed in the last section is easy to understand and use. However it has some disadvantages. First, we do not have precise control over the passband and stopband edge frequencies. Second, and the most important, the approximation error or ripples are not uniformly distributed over both passband and stopband intervals. This error is higher near the band edges and lower in regions away from band edges. By distributing approximation error uniformly over both passband and stopband, one can obtain lower order filter satisfying the same specifications.

Using the design algorithm due to Parks and McClellan [9], it is possible to design FIR filters that are optimum in approximating magnitude specifications with minimum peak error for the given order or minimum order for the given error. This algorithm is based on the fact that the frequency response for a linear phase FIR filter can be expressed as a polynomial in $\cos(\omega)$. The problem can then be changed to one of Chebyshev approximation where the best approximation to the given magnitude response is known to have equiripple behavior. The essence of this algorithm is to determine a polynomial solution so that the maximum value of the approximation error is minimized. The Parks-McClellan algorithm incorporates the well-known Remez-exchange routine for polynomial solution. The details of this algorithm are provided in many textbooks on digital signal processing including in [8]. A computer program written by Parks and McClellan [9] is also available for designing linear phase FIR filters which is based on their algorithm. Using this program we give designs of FIR filters described in the first two examples. Our purpose here is to compare design performance in terms of filter order and stopband attenuation.

Example 6-5:

Consider the lowpass filter specifications given in example 6-1, which are repeated below.

Sampling frequency:	8 KHz
Passband:	0 – 1 KHz
Stopband:	1.4 – 4 KHz
Stopband Attenuation:	50 DB

When this filter was designed using the Parks-McClellan algorithm, the filter length of 55 was obtained for stopband attenuation of 50 DB. The filter coefficients are given in Table 6-7 and the frequency response is shown in Figure 6-5.

```
                        Filter Length = 55
    h( 0) =   9.803186E-04 = h(54)        h(14) =  -4.902462E-03 = h(40)
    h( 1) =  -1.600514E-03 = h(53)        h(15) =  -1.769491E-02 = h(39)
    h( 2) =  -2.294067E-03 = h(52)        h(16) =  -1.748833E-02 = h(38)
    h( 3) =  -1.629659E-03 = h(51)        h(17) =  -9.733222E-05 = h(37)
    h( 4) =   8.873159E-04 = h(50)        h(18) =   2.348039E-02 = h(36)
    h( 5) =   3.600590E-03 = h(49)        h(19) =   3.248904E-02 = h(35)
    h( 6) =   3.744131E-03 = h(48)        h(20) =   1.259944E-02 = h(34)
    h( 7) =   5.530444E-05 = h(47)        h(21) =  -2.853314E-02 = h(33)
    h( 8) =  -5.215559E-03 = h(46)        h(22) =  -5.999381E-02 = h(32)
    h( 9) =  -7.280297E-03 = h(45)        h(23) =  -4.512416E-02 = h(31)
    h(10) =  -2.825630E-03 = h(44)        h(24) =   3.199079E-02 = h(30)
    h(11) =   6.049525E-03 = h(43)        h(25) =   1.499068E-01 = h(29)
    h(12) =   1.206036E-02 = h(42)        h(26) =   2.569771E-01 = h(28)
    h(13) =   8.270248E-03 = h(41)        h(27) =   3.001142E-01 = h(27)
```

Table 6-7: Example 6-5 Filter Coefficients ⸺⸺⸺⸺⸺

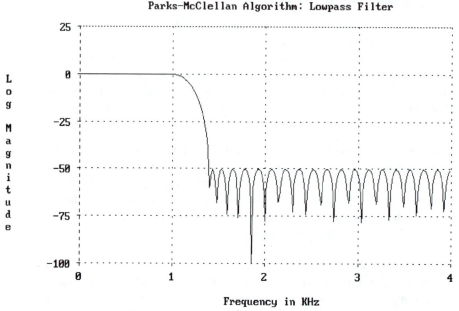

Figure 6-5: Example 6-5 Frequency Response

Example 6-6:

Consider the bandpass filter specifications given in example 6-2, which are repeated below.

Sampling frequency:	8 KHz
Lower Stopband:	0 – 0.6 KHz
Passband:	1 – 2 KHz
Upper Stopband:	2.4 – 4 KHz
Stopband Attenuation:	40 DB
Window Function:	Hanning Window

When this filter was designed using the Parks-McClellan algorithm, the filter length of 55 was obtained for stopband attenuation of 40 DB. The filter coefficients are given in Table 6-8 and the frequency response is shown in Figure 6-6.

```
                       Filter_Length_=_42
h( 0) = -5.160148E-03 = h(41)    |  h(11) = -1.004880E-02 = h(30)
h( 1) =  4.708251E-03 = h(40)    |  h(12) =  4.941909E-02 = h(29)
h( 2) =  7.186445E-03 = h(39)    |  h(13) =  5.083727E-02 = h(28)
h( 3) =  1.529386E-03 = h(38)    |  h(14) = -6.520151E-03 = h(27)
h( 4) =  5.241538E-03 = h(37)    |  h(15) =  1.120198E-02 = h(26)
h( 5) =  1.259914E-02 = h(36)    |  h(16) =  4.585057E-02 = h(25)
h( 6) = -4.231441E-03 = h(35)    |  h(17) = -9.028542E-02 = h(24)
h( 7) = -2.406552E-02 = h(34)    |  h(18) = -2.406141E-01 = h(23)
h( 8) = -9.823139E-03 = h(33)    |  h(19) = -6.130626E-02 = h(22)
h( 9) =  1.702993E-03 = h(32)    |  h(20) =  2.873955E-01 = h(21)
h(10) = -2.046206E-02 = h(31)    |
```

Table 6-8: Example 6-6 Filter Coefficients ——————————————

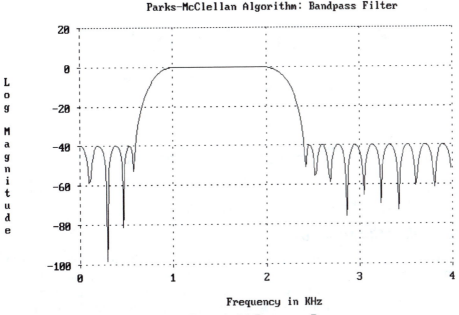

Figure 6-6: Example 6-6 Frequency Response

After comparing these designs with the corresponding window designs, it is obvious that the Parks-McClellan algorithms gives the smallest filter order and equiripple stopband performance. The smallest filter order is desirable from an implementation point of view. However in an ADSP-2101 implementation, we do not pay any penalty for a larger filter order so long as it is moderate. On the other hand, the window design technique allows us to design FIR filters using simple design equations without resorting to any sophisticated programs. This will allow us to concentrate on implementations and programming the ADSP-2101. In practice, when a more refined filter design is required, we can change the assembly language program by updating the filter coefficients obtained from the Parks-McClellan algorithm.

6.3 SINGLE-PRECISION FIR DIRECT FORM FILTER

The realization of an FIR filter can take many forms, although the most useful in practice are the Direct Form (or transversal filter) and the lattice structures. The FIR lattice filter is described later in Section 6.5. In this section we describe the realization of a single precision FIR Direct Form structure. This structure can be obtained directly from equation (6-1) for discrete-time convolution which is repeated below.

$$y(n) \; = \; \sum_{k=0}^{N-1} h(k)x(n-k)$$

We have already studied this equation and its implementation in section 5.4. A graphic representation of this Direct Form structure is shown in Figure 6-7.

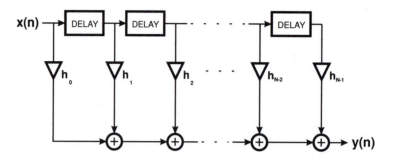

Figure 6-7: FIR Filter - Direct Form Structure

6.3.1 A Filter Subroutine

The subroutine that realizes this structure is shown in Listing 6-1. The first instruction sets up the computation by clearing MR and loading MX0 and MY0 with the first data and coefficient values from data and program memory. The multiply/accumulate with dual data fetch in the *conv* loop is then executed $N - 1$ times in N cycles to compute the sum of the first $N - 1$ products. The final multiply/accumulate instruction is performed with the rounding mode enabled to round the result to the upper 24 bits of MR. MR1 is then conditionally saturated to its most positive or negative value based on the status of the overflow flag MV. In this manner, results are accumulated to the full 40 bit resolution of MR, with saturation of the output only if the final result overflowed beyond the least significant bits of MR.

```
{ Single Precision FIR Direct Form Filter Subroutine          FIRDIRFM.DSP

    Calling Parameters
        I0 -> Oldest Input Data Value, x(n-N+1), in Delay Line
        L0 = Filter Length (N) or Taps
        I4 -> Beginning of Filter Coefficient Table
        L4 = Filter Length (N) or Taps
        M0, M4 = 1
        CNTR = Filter Length - 1 (N - 1)

    Return Values
        MR1 = Filter Output y(n) (rounded and saturated)
        I0 -> Oldest Input Data Value in Delay Line
        I4 -> Beginning of Filter Coefficient Table

    Altered Registers
        MX0, MY0, MR

    Computation Time
        N - 1 + 5 + 2 = N + 6  Cycles

    All coefficients and data values are assumed to be in 1.15 format.
}
.MODULE firdirfm_sub;
.ENTRY fir_df;
```

```
fir_df:   MR=0, MX0=DM(I0,M0), MY0=PM(I4,M4);
          DO conv UNTIL CE;
conv:         MR=MR_MX0*MY0(SS), MX0=DM(I0,M0), MY0=PM(I4,M4);
          MR=MR+MX0*MY0(RND);
          IF MV SAT MR;
          RTS;

.ENDMOD;
```
——————————— **Listing 6-1: Single Precision FIR Direct Form Subroutine** ———————————

The limit on the number of filter taps attainable for a real-time implementation of the above filter routine is determined primarily by the processor cycle time, the sampling rate, and the number of other computations required. The FIR Direct Form routine given above requires a total of $N + 6$ cycles for a filter of length N; at an 8 KHz sampling rate and an instruction cycle time of 80 nanoseconds, this permits a filter of 1,400 taps with 150 instruction cycles for other operations.

6.3.2 An Example Program

As an application of the above subroutine, we now provide a complete example program to implement the lowpass FIR filter designed in Example 6-1. This program is shown in Listing 6-2. The filter coefficients are given in Table 6-3 and the frequency response is shown in Figure 6-1. These coefficients are available in the disk file FIRDFLP.DAT as 1.15 hexadecimal format numbers.

This example program (and other programs to follow) uses a subroutine *CntlReg_inits* to initialize all control registers as discussed in Chapter 5. This subroutine is also available in the disk file CREGINIT.DSP. Assemble and link the main program and two subroutines to create an executable file. Load this executable file in the emulator and experiment with speech as well as other signals from a signal generator.

```
{ Lowpass FIR Filter                                        FIRDFLP.DSP
        Using Single Precision Direct Form Structure

    This program implements a single precision Direct Form FIR filter
    structure using subroutine fir_df available in disk file FIRDIRFM.DSP.
    The filter coefficients used in this program are for a lowpass filter
    of length 81 with passband from 0 to .125 and stopband from .15 to .5.
    It is designed via Parks-McClellan algorithm.  These coefficients are
    stored in a disk file, FIRDFLP.DAT.

    This program is written for EZ-ICE and EZ-LAB system with EZLAB1.ACH
    architecture file.  Additionally, this program uses subroutine
    CntlReg_inits to initialize all control registers.  It is available in
    disk file CREGINIT.DSP.  Assemble FIRDFLP.DSP, FIRDIRFM.DSP, and
    CREGINIT.DSP using ASM21.EXE.  Link using LD21.EXE to produce
    FIRDFLP.EXE.  Load FIRDFLP.EXE in EZ-ICE and execute.
}
.MODULE/RAM/ABS=0/BOOT=0              FIR_DFLP;
.PORT              write_dac0;
.PORT              load_dac;
.CONST             taps=81;
.VAR/CIRC          data[taps];
.VAR/PM/CIRC       fir_coeff[taps];
.INIT              fir_coeff: <firdflp.dat>;
    JUMP start; RTI; RTI; RTI;        {Reset Vector}
    RTI; RTI; RTI; RTI;               {irq2}
```

```
       RTI; RTI; RTI; RTI;                {sport0 TX}
       JUMP sample; RTI; RTI; RTI;        {sport0 RX}
       RTI; RTI; RTI; RTI;                {irq0}
       RTI; RTI; RTI; RTI;                {irq1}
       RTI; RTI; RTI; RTI;                {timer}
{--- Initialize ---------------------------------------------------------}
start:    CALL CntlReg_inits;            { set up SPORTS, TIMER, etc }
          I0=^data;      M0=1; L0=taps;
          I4=^fir_coeff; M4=1; L4=taps;
          CNTR=taps;
          DO zero UNTIL CE;
zero:     DM(I0,M0)=0;                    {clear the filter delay line buffer}
          ICNTL=B#00111;        { disable IRQ nesting, all IRQs edge-senstv}
          IMASK=B#101000;       { enable IRQ2 and SPORT0_RX interrupt }
wait:     IDLE;
          JUMP wait;
{--- Process Input Sample -----------------------------------------------}
sample:   SR1=RX0;                        {get new sample from microphone }
          DM(I0,M0)=SR1;                  {store sample in data buffer }
          CNTR=TAPS-1;
          CALL FIR_DF;                    {Filter current data}
          TX0=MR1;                        {filtered output to SPORT (to spkr) }
          SR = ASHIFT MR1 BY 1 (HI);
          DM(WRITE_DAC0)=SR1;             {latch sample for D/A }
          DM(LOAD_DAC)=SR1;               {display sample on scope via D/A}
          RTI;

.ENDMOD;
```

——————————— Listing 6-2: **Lowpass Filter Program** ———————————

6.3.3 Exercises

1. Design a highpass FIR filter using the window design technique to approximate the following specifications:

Sampling frequency:	8 KHz
Stopband:	0 – 1.6 KHz
Passband:	2 – 4 KHz
Stopband Attenuation:	50 DB
Window Function:	Hamming Window

    Implement this filter in the emulator and verify its operation.

2. Consider the FIR differentiator designed in Example 6-3. Create a disk file for the filter coefficients using 1.15 hexadecimal format numbers. Implement this differentiator in the emulator and study its effects on speech signals.

6.4 DOUBLE-PRECISION FIR DIRECT FORM FILTER

Many digital filters require a sum-of-products computation using operands that are greater than 16 bits in magnitude. The subroutine, DFIRDIRF.DSP, described in this section implements a sum-of-products calculations using coefficients and data that are both represented in double precision (or 32 bits). On the ADSP-2101, this is accomplished through the use of the mixed-mode multiply instructions. We provide this subroutine to illustrate programming methodology required in double-precision arithmetic.

6.4.1 A Filter Subroutine

The subroutine that realizes the double-precision sum-of-products operation used in computing the Direct Form filter is shown in Listing 6-3. First, the sum of products of the low halves of the coefficients and the high halves of the data values is computed; this sum is accumulated with the sum of the products of the high halves of the coefficients and the low halves of the data values. This sum is then shifted right 16 bits and then accumulated with the sum of the products of the high halves of the coefficients and the high halves of the data values. A conditional saturation is then performed on the final 32-bit result before storage to data memory. Note that because the result is only the most significant 32 bits, the products of the low-order coefficients and the low-order data affect only the least significant bit of the result and are therefore not computed.

```
{ Double Precision FIR Direct Form Filter Subroutine          DFIRDIRF.DSP

        Calling Parameters
                I0 --> Oldest input data value in delay line
                L0 = 2 * Filter length (N)
                I4 --> 2nd location (LSW of 1st value)
                       of filter coefficient table
                L4 = 2 * Filter length (N)
                M0,M4 = 1
                M1,M5 = 2
                M2,M6 = 3
                AX0 = Filter length - 2 (N-2)
                CNTR = Filter length - 2 (N-2)

        Return Values
                MR1,MR0 = sum of products
                (conditionally saturated to 32 bits)
                I0 --> Oldest input data value in delay line
                I4 --> 2nd location (LSW of 1st value)
                       of filter coefficient table

        Altered Registers
                MX0,MY0,MR

        Computation Time
                3 * (N - 2) + 16 + 9

        All coefficients and data values are assumed to be in 1.15 format.
}
.MODULE dfirdirfm_sub;
.ENTRY  dfir_df;

dfir_df:MR=0, MX0=DM(I0,M1), MY0=PM(I4,M5);
        DO hlloop UNTIL CE;
hlloop:         MR=MR+MX0*MY0(SU), MX0=DM(I0,M1), MY0=PM(I4,M5);
        MR=MR+MX0*MY0(SU), MX0=DM(I0,M2), MY0=PM(I4,M4);
        MR=MR+MX0*MY0(SU), MX0=DM(I0,M1), MY0=PM(I4,M5);
        CNTR=AX0;
        DO lhloop UNTIL CE;
lhloop:         MR=MR+MX0*MY0(US), MX0=DM(I0,M1), MY0=PM(I4,M5);
        MR=MR+MX0*MY0(US), MX0=DM(I0,M0), MY0=PM(I4,M5);
        MR=MR+MX0*MY0(US), MX0=DM(I0,M1), MY0=PM(I4,M5);
        MR0=MR1;            {downshift 16 places}
        MR1=MR2;
        CNTR=AX0;
        DO hhloop UNTIL CE;
hhloop:         MR=MR+MX0*MY0(SS), MX0=DM(I0,M1), MY0=PM(I4,M5);
```

```
        MR=MR+MX0*MY0(SS), MX0=DM(I0,M1), MY0=PM(I4,M6);
        MR=MR+MX0*MY0(SS);
        IF MV SAT MR;
        RTS;
.ENDMOD;
```
———————— **Listing 6-3: Double Precision FIR Direct Form Subroutine** ————————

6.4.2 Exercises

1. The above double-precision subroutine can be easily adapted to applications requiring mixed precision. For example, to use 32-bit coefficients and 16-bit data values, one would eliminate the *lhloop* and make the corresponding changes in the data memory pointer values and the size of the circular buffer. Modify the above subroutine to implement the above mixed precision arithmetic. Use the lowpass filter given in Table 6-1 to generate a filter coefficient file in 1.31 hexadecimal format. Using the subroutine and the filter coefficient file, implement the filter and verify its operation.

2. Combine both single- and mixed-precision FIR subroutines in a program to study coefficient quantization errors. Implement program lines to compute and output the error. Determine the error at various frequencies and plot the error response.

6.5 ALL-ZERO LATTICE FILTER

The lattice filter is extensively used in digital speech processing and in the implementation of adaptive filters. It is a preferred form of realization over other FIR or IIR filter structures because in speech analysis and in speech synthesis, the small number of coefficients allow a large number of "formants" to be modelled in real time. Its physical analogue is a series of cylinders of different radii; each of the filter coefficients represents the amount of energy reflected at a boundary of two cylinders. The all-zero lattice is the FIR representation of the lattice filter while the all-pole lattice is the IIR representation. The all-pole lattice filters are described in Chapter 7. In this section, we describe a subroutine to implement the all-zero lattice filter on the ADSP-2101 microcomputer.

An FIR filter of length N (or order $(N-1)$) has a lattice structure with $(N-1)$ stages as shown in Figure 6-8.

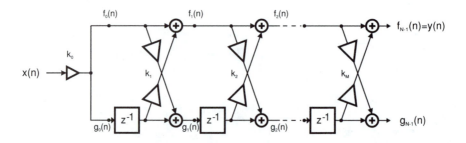

Figure 6-8: All-Zero Lattice Filter

Each stage of the filter has an input and output that are related by the order-recursive equations ([8])

$$f_m(n) = f_{m-1}(n) + k_m g_{m-1}(n-1) \qquad m = 1, 2, \ldots, N-1$$

$$g_m(n) = k_m f_{m-1}(n) + g_{m-1}(n-1) \qquad m = 1, 2, \ldots, N-1$$

where the parameters k_m, $m = 1, 2, \ldots, N-1$, called the *reflection coefficients*, are the lattice filter coefficients. If the initial values of $f_m(n)$ and $g_m(n)$ are both the scaled value of the filter input $x(n)$, then the output of the $(N-1)$ stage lattice filter corresponds to the output of an $(N-1)$ order FIR filter, that is,

$$f_0(n) = g_0(n) = k_0 x(n)$$

$$y(n) = f_{N-1}(n)$$

For example, if $N = 3$ (second-order FIR filter) then the lattice filter equations are:

$$f_1(n) = k_0 x(n) + k_1 k_0 x(n-1)$$

$$g_1(n) = k_1 k_0 x(n) + k_0 x(n-1)$$

$$f_2(n) = f_1(n) + k_2 g_1(n-1)$$

$$g_2(n) = k_2 f_1(n) + g_1(n-1)$$

If we focus our attention on $f_2(n)$ and substitute for $f_1(n)$ and $g_1(n-1)$ from the above equations, we obtain

$$y(n) = f_2(n) = k_0 x(n) + k_1 k_0 x(n-1) + k_2[k_1 k_0 x(n-1) + k_0 x(n-2)]$$

$$= k_0 x(n) + k_0 k_1 (1 + k_2) x(n-1) + k_0 k_2 x(n-2)$$

Now this equation is identical to the output of the direct form FIR filter given by

$$y(n) = h(0)x(n) + h(1)x(n-1) + h(2)x(n-2)$$

If we equate the coefficients, we have

$$h(0) = k_0 \qquad h(1) = k_0 k_1 (1 + k_2) \qquad h(2) = k_0 k_2$$

or, equivalently,

$$k_0 = h(0) \qquad k_2 = \frac{h(2)}{h(0)} \qquad k_1 = \frac{h(1)}{h(0) + h(2)}$$

Thus the reflection coefficients k_m, $m = 0,1,2$ can be obtained from the impulse response $h(n)$, $n = 0,1,2$ of the FIR filter. A similar analysis can be done to demonstrate the equivalence between an Nth-order direct form FIR and an N-stage lattice filter which is given in [8].

The ADSP-2101 implementation of the all-zero lattice filter is shown in Listing 6-4. This subroutine computes the entire output sequence given the input sequence of finite length stored in a buffer. Therefore this subroutine is different from the previous subroutines in which only one output sample was computed per subroutine call. There is also another important difference in the implementation. In the direct form structure, all the intermediate results were stored in the MR register which is 40-bits wide. Therefore, overflow was not an issue and only the final result was checked for overflow. In the lattice filter implementation, the intermediate signal values f_m's and g_m's cannot be stored in the MR register. These signal values must also be protected from an overflow since an addition is required in their calculation. However, a saturation logic may lead to a disaster since these values are used several times in the output calculations. Therefore the desirable approach is to provide sufficient dynamic range for calculations. This must be determined based on the filter coefficients and the filter order.

Before this subroutine is called, several registers must be pre-loaded. The index register I0 should contain the starting address of the input buffer, and I2 should hold the starting address of the output buffer. The index register I3 should contain the starting address of the filter delay line, and I4 should contain the starting address of the coefficient buffer. The length registers L0 and L2 should be set to 0, but L3 and L4 should be set to the order of the filter (or number of stages) to make the delay line and coefficient buffers circular. The modify register M3 should be set to one, and the SE register should contain the value needed to maintain a valid output data format for sufficient dynamic range. (For example, if two 1.15 numbers are multiplied, the product is a 1.31 number To obtain a product in 4.28 format, the SE register should be set to -3.) The multiplier feedback register, MF, should contain the value one in the output format. Multiplication by MF is an alternative method of converting the output to the correct format. The CNTR register should contain the number of locations in the output buffer (or the number of output samples to be computed).

The *out-lp* loop is executed once for each output data point. The CNTR register is loaded with the order of the filter, and the first input data point is loaded into MX0. The *latt_lp* loop performs the filtering operation on the input data point. The first multiplication in the *latt_lp* loop formats the $f_{m-1}(n)$ value into the MR register and also reads in values for $g_{m-1}(n-1)$ and k_m. These values are then multiplied and accumulated to produce $f_m(n)$, at the same time the value $g_{m-1}(n)$ is stored in the delay line for the next pass. The value $f_m(n)$ is reformatted in the shifter for use by the multiplier in the next pass of the *latt_lp* loop. Next, $g_{m-1}(n-1)$ is formatted into the multiplier to compute the value of $g_m(n)$. This value is then accumulated with the product of k_m and $f_m(n)$. Again, the shifter reformats the value before storage.

```
{ All-Zero Lattice Filter Subroutine                                    FIRLATFL.DSP

      Calling Parameters
          CNTR = Length of Output Buffer
          I0 --> Input Buffer                         L1 = 0
          I2 --> Output Buffer                        L2 = 0
```

```
               I3 --> Delay Line Buffer (circular)    L3 = Filter Order
               I4 --> Coefficient Buffer (circular)   L4 = Filter Order
               M0 = 1
               M2 = 0
               M3 = 1
               M4 = 1
               SE = Appropriate Scale Factor
               MF = Formatted 1

        Return Values
            Output Buffer Filled

        Altered Registers
            MX0,MX1,MY0,MF,MR,SR,I2,I3,I4

        All coefficients and data values are assumed to be in 1.15 format.

        Computation Time
            (8 * Filter Order + 4) * Output Buffer Length + 3 + 1 cycles
}

.MODULE firlatfl_sub;
.ENTRY fir_lf;

fir_lf:   SR1=0;                                   {Clear SR1 for first pass}
          DO out_loop UNTIL CE;                    {Loop output length}
           CNTR=L3;
           MX0=DM(I0,M0);                          {Get x(n)}
           DO latt_loop UNTIL CE;                  {Loop through filter}
             MR=MX0*MF (SS), MX1=DM(I3,M2), MY0=PM(I4,M4); {Get g, k}
             MR=MR+MX1*MY0 (SS), DM(I3,M3)=SR1; {Compute fm store g}
             SR=ASHIFT MR1 (HI);                   {Reformat fm}
             SR=SR OR LSHIFT MR0 (LO);
             MR=MX1*MF (SS);                       {Format gm-1}
             MX0=SR1, MR=MR+MX0*MY0 (SS);          {Compute gm and Hold fm}
             SR=ASHIFT MR1 (HI);                   {Reformat gm}
latt_lp:     SR=SR OR LSHIFT MR0 (LO);
out_lp:   DM(I2,M0)=MX0;                           {Save output}
          RTS;
.ENDMOD;
```

——————————— **Listing 6-4: All-Zero Lattice Filter Subroutine** ———————————

6.6 SINGLE SIDEBAND (SSB) MODULATOR

As an interesting project on FIR filtering, let us consider Single Sideband (SSB) modulation. This is a well known modulation technique in analog communication systems. We will simulate this technique in the ADSP-2101 system for speech signals. The project incorporates many earlier modules that we studied before, including a sinusoidal waveform generator, a Hilbert transformer, a delay-line, etc. Therefore in this project, we will write a program to implement a complete system and analyze its performance.

SSB modulation was developed to address two problems in Amplitude Modulation (AM), namely wasteful transmitted power and transmission bandwidth. Since the upper and lower sidebands of AM are uniquely related by symmetry about the carrier frequency, given the amplitude and phase of one we can always reconstruct the other. Suppressing the carrier and one sideband overcomes the shortcomings in AM. This leads to SSB modulation. There are several approaches to achieving SSB modulation. One approach which is appropriate for our purpose involves the use of a Hilbert transformer, which is a all-pass filter that imparts a 90° phase shift on the signal at its input. For an arbitrary signal $x(t)$, it can be shown that [10] the SSB modulated signal $x_c(t)$ is given by

$$x_c(t) = \frac{1}{2}[x(t)\cos(2\pi F_c t) \mp \hat{x}(t)\sin(2\pi F_c t)] \qquad (6-5)$$

where the upper sign is taken for upper SSB and vice versa, F_c is the carrier frequency, and where $\hat{x}(t)$ is the Hilbert Transform of $x(t)$. A block diagram of the SSB modulator based on the above equation is shown in Figure 6-9. We will now see how to implement this analog modulator using the ADSP-2101 DSP system.

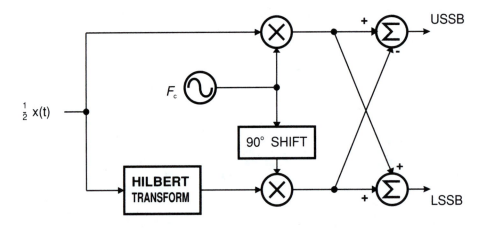

Figure 6-9: Single Sideband Modulator

6.6.1 A Frequency Shifter

When the upper sign is used in equation (6-5), we obtain the upper sideband of the signal shifted in frequency by F_c amount. In this mode, the SSB modulator functions as a frequency shifter. Listing 6-5 shows a program which shifts an speech signal by 2 KHz. It is available on the disk as FIRDFSSB.DSP file.

The frequency shifter implemented in this program uses the 51-tap FIR Hilbert transformer designed in Example 6-4. The filter is implemented using subroutine *fir_df* from Section 6.3.1 and it uses two circular buffers: *data* and *fir_coeff*. Due to the linear-phase characteristics, this filter imparts a 25-sample shift on the input signal. Therefore the input signal in the upper branch of Figure 6-9 must also be delayed by the same amount for proper synchronization. This is done by using a circular buffer called *sig_delay*. We need two digital oscillators, one for the cosine waveform and the other for the sine waveform. This can be achieved by using one oscillator along with a delay line as a phase shifter. A digital oscillator is implemented using the *sine* subroutine developed in Chapter 5. The circular buffer *cos_sin* performs the function of two oscillators. In this project, the sampling frequency is 8 KHz while the carrier frequency is 2 KHz. Therefore a delay of two samples provides a phase-shift of 90 °. This can be achieved by the *cos_sin* buffer of length 3. Finally, a circular buffer *angles* stores 8 angle values to generate a 2 KHz waveform. The rest of the program is self explanatory.

```
{ Single Sideband Modulator                               FIRDFSSB.DSP
           Using Single Precision Direct Form Structure

     This program implements a Single Sideband (SSB) Modulator using Hilbert
     Transform Approach.  The Hilbert Transformer is implemented using a
     51-tap single precision FIR filter.  The coefficients of this filter
     are available in FIRDFHT.DAT disk file.  The program also uses
     subroutine sin to generate cosine and sine components of carrier
     frequency at 2 KHz.  This subroutine is available in disk file
     SINE.DSP.

     This program is written for EZ-ICE and EZ-LAB system with EZLAB1.ACH
     architecture file.  Additionally, the program uses subroutine
     CntlReg_inits to initialize all control registers.  This subroutine is
     available in disk file CREGINIT.DSP.  Assemble FIRDFSSB.DSP, SIN.DSP,
     FIRDIRFM.DSP and CREGINIT.DSP using ASM21.EXE.  Link using LD21.EXE to
     produce FIRDFSSB.EXE.  Load FIRDFSSB.EXE in EZ-ICE and execute.
}
.MODULE/RAM/ABS=0/BOOT=0                   FIR_SSBM;
.CONST              taps=51;
.CONST              alpha=25;
.VAR/DM/CIRC        data[taps];
.VAR/PM/CIRC        fir_coeff[taps];
.VAR/DM/CIRC        sig_delay[alpha];
.VAR/PM/CIRC        angles[8];
.VAR/DM/CIRC        cos_sin[3];
.INIT              fir_coeff: <firdfht.dat>;
.INIT              angles: 0x000000, 0x200000, 0x400000, 0x600000,
                           0x800000, 0xA00000, 0xC00000, 0xE00000;
{-------------------------------------------------------------------------}
    JUMP START; RTI; RTI; RTI;            {Reset Vector}
    RTI; RTI; RTI; RTI;                   {Irq2}
    RTI; RTI; RTI; RTI;                   {Sport0 TX}
    JUMP SAMPLE; RTI; RTI; RTI;           {Sport0 RX}
    RTI; RTI; RTI; RTI;                   {Irq0}
    RTI; RTI; RTI; RTI;                   {Irq1}
    RTI; RTI; RTI; RTI;                   {Timer}
{--- Initialize ----------------------------------------------------------}
start:    CALL CntlReg_inits;            { set up SPORTS, TIMER, etc }
          I0=^data;       M0=1; L0=taps;
          I1=^sig_delay; M1=0; L1=alpha;
          I2=^cos_sin;          L2=2;
                          M3=1; L3=0;
          I4=^fir_coeff; M4=1; L4=taps;
          I5=^angles;    M5=0; L5=8;
          CNTR=taps;
          DO zero1 UNTIL CE;
zero1:    DM(I0,M0)=0;                    {Clear the filter delay line buffer}
          CNTR=alpha;
          DO zero2 UNTIL CE;
zero2:    DM(I1,M0)=0;                    {Clear the signal delay line buffer}
          ICNTL=b#00111;         { disable IRQ nesting, all IRQs edge-senstv}
          IMASK=b#101000;        { enable IRQ2 and SPORT0_RX interrupt }
wait:     IDLE;
          JUMP wait;
{--- Process Input Sample ------------------------------------------------}
sample:   SR1=RX0;                        {Get new sample from microphone }
          DM(I0,M0)=SR1;                  {Store sample in data buffer }
          MY0=DM(I1,M1);                  {In phase component of signal in MY0}
          DM(I1,M0)=SR1;                  {Store sample in sig_delay buffer}
```

```
        AX0=DM(I5,M4);                   {Angle from angles buffer in AX0}
        CALL sin;                        {Cosine value in AR}
        DM(I2,M0)=AR;                    {Store cosine value in cos_sin buffer}
        MR1=AR*MY0;                      {product of cos and in-phase in MR1}
        SR=ASHIFT MR1 BY -1 (HI);        {scale by 0.5}
        AX0=SR1;                         {scaled product in AX0}
        CNTR=taps-1;
        CALL fir_df;                     {Quadrature component in MR1}
        MY0=DM(I2,M1);                   {Sin value in MY0}
        MR1=MR1*MY0;                     {product of sin and quad in MR1}
        SR=ASHIFT MR1 BY -1 (HI);        {Scale by 0.5}
        AY0=SR1;                         {Move to AY0}
        AR=AX0-AY0;                      {Upper SSB modulated signal in AR}
        IF AV SAT AR;
        TX0=AR;                          {modulated signal to SPORT (to spkr) }
        RTI;
{------------------------------------------------------------------------}
.ENDMOD;
```
———————— Listing 6-5: **A Frequency Shifter Program** ————————————

    Assemble, link, and execute this program in the emulator. Be careful to use proper
architecture and filter coefficient files. Connect a function generator to the microphone port
and apply a tone of 1 KHz. Observe the output on a scope and verify that it is a 3 KHz tone.
Experiment with speech signals and characterize their output from the speaker.

6.6.2 Exercises

1. The above SSB modulator shifts the input signal, which has bandwidth up to 3.4 KHz
    (due to anti-aliasing filter in the codec), by 2 KHz. The frequency shifted output
    signal therefore will be distorted due to aliasing above 4 KHz (at an 8 KHz sampling
    rate). We should have implemented a lowpass filter to restrict frequency components
    below 2 KHz. Modify the above program to first include a lowpass filter with cutoff
    frequency of 1 KHz. Choose the proper specifications for the lowpass FIR filter.
    Also modify the program so that the upper SSB is available at DAC0 and the lower
    SSB is available at DAC1. Experiment with speech signals and listen to the outputs
    from both SSBs. Is the output from the lower SSB intelligible? If not, why not?

2. A single Sideband modulator can also be implemented by first multiplying the input
    signals by a carrier (which is AM) and then extracting the upper or lower SSBs using
    a sharp highpass or lowpass filter, respectively. Write a program to simulate a
    frequency shifter based on this idea. Prefilter the input signals up to 1 KHz. Use a
    carrier frequency of 2 KHz and a stopband for FIR filters beginning at 2 KHz.
    Compare the performance of this shifter with the one above.

6.7 SUMMARY

Design and implementation of FIR digital filters is a very important operation in DSP. In this
chapter, we first briefly discussed two useful design techniques. The window design is a simple
technique while the Parks-McClellan algorithm is an optimum one. Using simple programs
developed in Chapter 5, we then discussed implementation of FIR filters using single- and
double-precision arithmetic. We also introduced another FIR filter structure called the all-zero
lattice filter and described its implementation. Finally as an exercise in a complete system
design and implementation, we developed a SSB modulator as a frequency shifter. In the next
chapter, we will provide a similar treatment for IIR digital filters.

chapter 7

IIR FILTER IMPLEMENTATIONS

7.1 INTRODUCTION

In this chapter, Infinite-duration Impulse Response (IIR) filters are designed and implemented on the ADSP-2101 microcomputer. Compared to FIR filters, IIR filters can often be much more efficient in terms of attaining better magnitude response with a given filter order. This is because IIR filters incorporate feedback and are capable of realizing both poles and zeros of a system function, whereas FIR filters are only capable of realizing the zeros. This means that IIR filters can run faster and hence must be considered in applications where speed is important. They, however, have stability problems and have nonlinear phase characteristics which might make them unsuitable for some applications. FIR filters, on the other hand, are always stable and can be designed to have exact linear phase.

When it comes to implementation, there are more trade-offs that must be considered when choosing between FIR and IIR filters. As discussed in Chapter 6, FIR filters are generally implemented using a linear convolution. This implementation does not pose any problems on finite word-length processors, such as the ADSP-2101, because all filter coefficients can be suitably scaled to avoid saturation and overflow. IIR filters are implemented using difference equations as described in Chapter 5. One approach is to compute two convolution sums as two FIR filters over the past few inputs and outputs and take the difference to obtain the next output value. This approach has stability problems due to coefficient quantization and if the order is large then the filter actually can become unstable. Therefore in most cases, IIR filters are factored into second-order sections to minimize this sensitivity, and then the entire filter is implemented as a cascade or parallel network of these sections. Still each factor must be carefully implemented to avoid overall numerical overflow problems. There are other problems which stem from the fact that this filter is recursive. Even if the input is zero, the filter can enter into a bad mode in which there is some output called limit cycles. These and other implementation issues will be discussed in this chapter.

In the first half of the chapter, we briefly discuss the realization and design issues of IIR filters. We then present computer programs for three important structures on the ADSP-2101 microcomputer with emphasis on trouble-free implementations.

7.2 IIR FILTER STRUCTURES

Infinite-duration Impulse Response filters form a wider class of filters whose z-domain system function $H(z)$ is a rational function in z. This class includes FIR filters as a special case. However, we treated FIR filters separately for design as well as for implementation purposes. The system function of an IIR filter is given by

$$H(z) = \frac{Y(z)}{X(z)} = \frac{\sum\limits_{n=0}^{N} b_n z^{-n}}{\sum\limits_{n=0}^{N} a_n z^{-n}} = \frac{b_0 + b_1 z^{-1} + \dots + b_N z^{-N}}{1 - a_1 z^{-1} - \dots - a_N z^{-N}} \qquad (7-1)$$

where b_n and a_n are the coefficients of the filter. We have assumed that both the numerator and denominator polynomials are of order N. The order of such an IIR filter is called N if $a_N \neq 0$. The difference equation representation of an IIR filter was discussed earlier in Chapter 5. It is expressed as:

$$y(n) = \sum\limits_{m=0}^{N} b_m x(n-m) + \sum\limits_{m=1}^{N} a_m y(n-m) \qquad (7-2)$$

This representation implies that the output of an IIR filter is a function of the past outputs and the present and past inputs. This feedback (or recursive) nature of the IIR filter is a major challenge in its implementation especially on a digital signal processor as we shall see.

There are three different structures than can be used to implement an IIR filter.

- **Direct Form:** In this form, the difference equation (7-2) is implemented directly as given. Since there are two parts to this filter, namely FIR and recursive (or equivalently, the numerator and denominator parts), this implementation leads to two versions, Direct Form I and Direct Form II structures.

- **Cascade Form:** In this form, the system function $H(z)$ in equation (7-1) is factored into smaller second-order section, called *biquads*. The system function is then represented as a *product* of these biquads. Each biquad is implemented in a direct form and the entire system function is implemented as a *cascade* of biquad sections.

- **Parallel Form:** This is similar to the cascade form but after factorization, a partial fraction expansion is used to represent $H(z)$ as a *sum* of smaller second-order sections. Each section is again implemented in a direct form and the entire system function is implemented as a *parallel* network of sections.

We will briefly discuss the direct and cascade form in this section. The parallel form, due to its design, makes sense when more processors are available to implement all sections simultaneously. In this book, we will not discuss multiprocessor implementation. Therefore, the parallel form is of little use in our IIR implementations.

7.2.1 Direct Form

As the name suggests, the difference equation (7-2) is implemented as given using memory, multipliers and adders. For the purpose of illustration, let $N = 3$. Then the difference equation is

$$y(n) \;=\; b_0 x(n) + b_1 x(n-1) + b_2 x(n-2) + b_3 x(n-3) + b_4 x(n-4)$$

$$+ a_1 y(n-1) + a_2 y(n-2) + a_3 y(n-3) + a_4 y(n-4)$$

which can be implemented as shown in Figure 7-1. This block diagram is called *Direct Form I* structure.

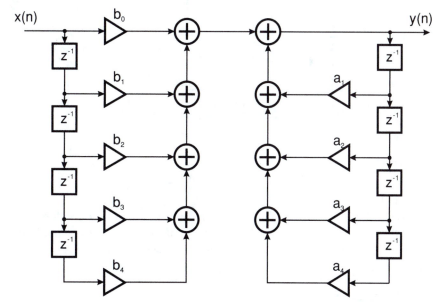

Figure 7-1: Direct Form I Structure

The direct form I structure implements each part of the rational function $H(z)$ separately with a cascade connection between them. The numerator, or FIR part, is a tapped delay line followed by the denominator, or recursive part, which is a feedback tapped delay line. Thus, there are two separate delay lines in this structure and hence it requires 8 memory elements. We can reduce this memory count or eliminate one delay line by interchanging the order in which the two parts are connected in the cascade. Now the two delay lines are close to each other, connected by a unity gain branch. Therefore one delay line can be removed and this reduction leads to a canonical structure called *Direct Form II* structure. It is shown in Figure 7-2. It should be noted that both direct forms are equivalent from the input-output point of view. Internally, however, they have different signals.

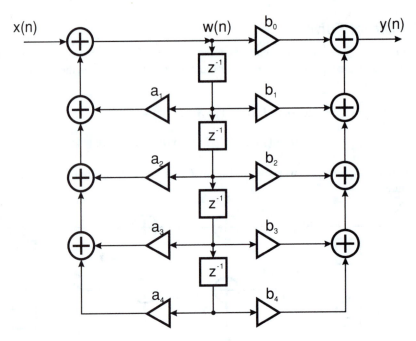

Figure 7-2: Direct Form II Structure

In section 7-4, we will mainly discuss the implementation of the direct form II structure on the ADSP-2101 because it is more efficient in terms of memory usage and speed. In Chapter 5, we briefly discussed direct form I implementation under the difference equation implementation topic.

7.2.2 Cascade Form

In this form, the system function $H(z)$ is written as a product of second order sections. In the remainder of this chapter, we assume that N is an even number. Then

$$
\begin{aligned}
H(z) &= \frac{b_0 + b_1 z^{-1} + \cdots + b_N z^{-N}}{1 - a_1 z^{-1} - \cdots - a_N x^{-N}} \\[2mm]
&= b_0 \frac{1 + \frac{b_1}{b_0} z^{-1} + \cdots + \frac{b_N}{b_0} z^{-N}}{1 - a_1 z^{-1} - \cdots - a_N x^{-N}} \\[2mm]
&= b_0 \prod_{k=1}^{M} \frac{1 + B_{1k} z^{-1} + B_{2k} z^{-2}}{1 - A_{1k} z^{-1} - A_{2k} z^{-2}}
\end{aligned}
\qquad (7-3)
$$

where M is equal to $\frac{N}{2}$, and B_{1k}, B_{2k}, A_{1k}, and A_{2k} are real numbers which are the coefficients of second order sections. The second order section

$$H_k(z) = \frac{Y_{k+1}(z)}{Y_k(z)} = \frac{1 + B_{1k}z^{-1} + B_{2k}z^{-2}}{1 - A_{1k}z^{-1} - A_{2k}z^{-2}} \quad k = 0, \ldots, M$$

with

$$Y_1(z) = b_0 X(z), \qquad Y_{M+1}(z) = Y(z)$$

is called the kth biquad section. The input to the kth biquad section is the output from the $(k-1)^{th}$ biquad section while the output from the kth biquad is the input to the $(k+1)^{th}$ biquad. Now each biquad section $H_k(z)$ can be implemented in direct form II as shown in Figure 7-3. The entire filter is then implemented as a cascade of biquads.

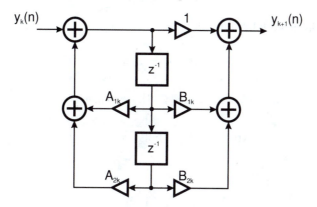

Figure 7-3: Biquad Section Structure

As an example, consider $N = 4$. Figure 7-4 shows a cascade form structure for this 4th order IIR filter.

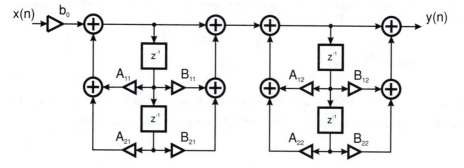

Figure 7-4: Cascade Form Structure for $N = 4$

In section 7-5, we will discuss the implementation of the cascade form structure on the ADSP-2101. From a finite word-length point of view, this implementation has advantages over the direct form as we shall see. But first, we briefly discuss the design of IIR filters in the next section.

7.2.3 Lattice Structure

In Section 6.5, we described a lattice filter structure that is equivalent to an FIR filter. In this section we present a similar structure for all-pole IIR filters. The implementation of IIR filters that contain both poles and zeros requires lattice-ladder structure which is not described here.

Consider an all-pole filter described by the system function

$$H(z) = \frac{b_0}{1 + \sum_{m=1}^{N} a_m z^{-k}}$$

The difference equation for this filter is

$$y(n) = b_0 x(n) - \sum_{m=1}^{N} a_m y(n-m) \qquad (7-4)$$

If we interchange $x(n)$ with $y(n)$ in equation (7-4), we obtain

$$x(n) = b_0 y(n) - \sum_{m=1}^{N} a_m x(n-m)$$

or, equivalently,

$$y(n) = \frac{1}{b_0} x(n) + \frac{1}{b_0} \sum_{m=1}^{N} a_m x(n-m)$$

We note that the above equation describes an all-zero (FIR) filter for which we have presented a lattice structure in Section 6.5. The lattice structure for the all-zero filter can now be used to obtain a lattice structure for an all-pole IIR filter by interchanging the roles of the input and output. We take the all-zero lattice filter illustrated in Figure 6-8 and redefine the input as

$$x(n) = f_N(n)$$

and the output as

$$y(n) = b_0 f_0(n)$$

These are exactly the opposite of the definitions for the all-zero lattice. The quantities $\{f_m(n)\}$ likewise are also computed in opposite order and the resulting set of equations is given by ([8])

$$f_N(n) = x(n) \qquad (7-5a)$$

$$f_{m-1}(n) = f_m(n) - k_m g_{m-1}(n-1) \qquad m = N, N-1, \ldots, 1 \qquad (7-5b)$$

$$g_m(n) = k_m f_{m-1}(n) + g_{m-1}(n-1) \qquad m = N, N-1, \ldots, 1 \qquad (7-5c)$$

$$y(n) = b_0 f_0(n) = b_0 g_0(n) \qquad (7-5d)$$

which corresponds to the structure shown in Figure 7-5.

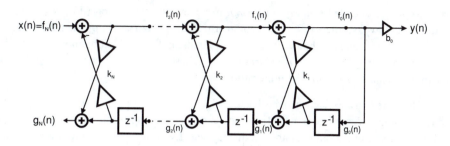

Figure 7-5: Lattice Filter Structure

To demonstrate that the set of equations (7-5) represent an all-pole IIR filter given in equation (7-4), consider the case in which $N = 2$. The equations above reduce to

$$f_2(n) \;=\; x(n)$$

$$f_1(n) \;=\; f_2(n) - k_2 g_1(n-1)$$

$$g_2(n) \;=\; k_2 f_1(n) + g_1(n-1)$$

$$f_0(n) \;=\; f_1(n) - k_1 g_0(n-1)$$

$$g_1(n) \;=\; k_1 f_0(n) + g_0(n-1)$$

$$y(n) \;=\; b_0 f_0(n) = b_0 g_0(n)$$

After some simple substitution and manipulations we obtain

$$y(n) \;=\; b_0 x(n) - b_0 k_1(1+k_2)y(n-1) - b_0 k_2 y(n-2) \tag{7-6}$$

Comparing the equation (7-4) with (7-6), we note that the two representations are equivalent if

$$a_1 \;=\; b_0 k_1(1+k_2) \qquad a_2 \;=\; b_0 k_2$$

A similar analysis can be done to establish the equivalence between an Nth-order direct form all-pole IIR filter and an N-stage all-pole lattice filter ([8]).

In Section 7-6, we will discuss the implementation of the all-pole lattice filter on the ADSP-2101.

7.3 OVERVIEW OF IIR FILTER DESIGN

IIR filters have long (infinite-duration) impulse responses. Hence they can be matched to analog filters, all of which generally have infinitely long impulse responses. Therefore the basic IIR filter design technique transforms well known analog filters into digital filters using complex-valued mappings. The advantage of this technique lies in the fact that both analog filter design tables and complex-valued mappings are extensively available in literature. This basic technique is called *Filter Transformation*. However the analog filter design tables are available for lowpass filters only. One would also like to design and implement other frequency selective filters such as highpass, bandpass, or bandstop filters. To obtain these other filters, one needs to apply *Spectral Transformations*. These transformations are complex-valued mappings that are also available in literature.

There are two approaches to this basic design technique in which a digital filter is obtained from an analog lowpass filter. In the first approach, an analog spectral transformation is applied to the lowpass filter to obtain another frequency selective filter. Then a filter transformation is applied to obtain the desired digital filter. In the second approach, first a lowpass digital filter is obtained from the corresponding analog filter through a filter transformation. Then the desired frequency selective digital filter is obtained using a filter-band transformation. In this book we will explain the second approach, while in most software packages on filter design the first approach is used.

Given the frequency domain filter specifications, the essence of IIR filter design is to obtain the filter of order N, and the filter coefficients a_k, b_k (for direct form structures) or A_{ik}, B_{ik} (for cascade form structure). Therefore the following steps are required in this design:

- Using the given digital specifications, obtain the corresponding lowpass analog filter specifications. This step requires inverse filter and spectral transformations on critical band or cutoff frequencies.

- Design an analog lowpass filter. This step requires use of well known analog filter design tables or formulas.

- Apply the filter transformations to obtain a digital lowpass filter. There are two useful transformations available; impulse invariance and bilinear, out of which the later one is a better choice.

- Apply a spectral transformation to obtain the desired digital filter from the lowpass filter. The result of this step is the system function $H(z)$ in the form of equation (7-1).

- Finally, if the cascade form structure is desired then factor $H(z)$ to obtain biquad sections in the form of equation (7-3).

The main problem with this technique is that we have no control over the phase response which ideally should be linear in the passband for most applications. This is because analog filter design equations are based on magnitude-squared frequency response. For the precise linear-phase response, one must consider FIR filters discussed in Chapter 6.

7.3.1 Analog Lowpass Filter Design

The IIR filter design techniques that we are discussing rely on existing analog lowpass filters (LPF) to obtain the desired digital filters. We will call these analog filters *prototype filters*. There are three prototypes which are widely used in practice. These are: Butterworth LPF, Chebyshev LPF, and Elliptic LPF. Butterworth filters have a flat response in both passband and stopband, while Chebyshev filters have either equiripple passband and flat stopband (Type-I) or flat passband and equiripple stopband (Type-II). In this section, we will briefly summarize design equations to obtain system functions of Butterworth and Chebyshev Type-I filters given the frequency domain specifications. The elliptic filters exhibit equiripple behavior in both the passband and stopband, and hence they are optimum in that they achieve the minimum order N for the given specifications. However, they are very difficult to analyze. It is not possible to design them using simple tools and often computer programs are needed to design them. Therefore we will not discuss them in this book since the lower order of the elliptic filter is not an important issue in our implementations.

Analog filters are specified in terms of their frequency-domain transfer function given by

$$H_a(j\Omega) \;=\; \frac{Y_a(j\Omega)}{X_a(J\Omega)}$$

where subscript a denotes an analog quantity, and Ω is the analog frequency. Analog prototype filters are characterized in terms of their magnitude-squared response given by

$$|H_a(j\Omega)|^2 \;=\; H_a(j\Omega)H_a(-j\Omega) \;=\; H_a(s)H_a(-s)\big|_{s=j\Omega}$$

where $H_a(s)$ is the s-domain system function. Thus,

$$H_a(s)H_a(-s) = |H_a(j\Omega)|^2\big|_{\Omega=-js} \qquad\qquad (7-7)$$

Therefore the poles and zeros of the magnitude-squared system function are distributed in a mirror-image symmetry with respect to the $j\Omega$ axis in the s-plane. A *causal* and *stable* $H_a(s)$ can now be obtained from this distribution by assigning left-half plane poles and zeros of $H_a(s)H_a(-s)$ to $H_a(s)$. The resulting filter is called a *minimum-phase* filter.

The magnitude-squared specifications on prototype filters are given in terms of parameters shown in Figure 7-6. These parameters are:

- passband cutoff frequency Ω_P in rad/sec,
- stopband cutoff frequency Ω_S in rad/sec,
- passband ripple R_P in dB, and
- stopband attenuation A_S in dB

Given these parameters, we would like to determine $H_a(s)$ in order to obtain the prototype filter.

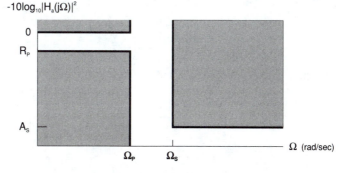

Figure 7-6: Analog Prototype Filter Specifications

Butterworth Filter Design

This filter is characterized by the property that its magnitude response is flat in both passband and stopband. The magnitude-squared response of an Nth-order lowpass filter is given by

$$|H_a(j\Omega)|^2 \quad = \quad \frac{1}{1 + \left(\frac{j\Omega}{j\Omega_c}\right)^{2N}} \tag{7-8}$$

where Ω_c is the 3-dB cutoff frequency. The system function $H_a(s)$ can be obtained from equation (7-6) by substituting $\Omega = -js$ to obtain

$$H_a(s)H_a(-s) \quad = \quad \frac{(j\Omega_c)^{2N}}{s^{2N} + (j\Omega_c)^{2N}}$$

The roots of the denominator polynomial or the poles of $H_a(s)H_a(-s)$ are given by

$$s_k \quad = \quad \Omega_c \exp\left[j\frac{\pi}{2N}(2k + N + 1)\right], k = 0, 1, \ldots, 2N - 1 \tag{7-9}$$

Now a stable and causal filter $H_a(s)$ can be obtained by selecting poles in the left-half plane resulting in the system function in the form:

$$H_a(s) \quad = \quad \frac{(\Omega_c)^N}{\prod\limits_{LHP\,poles}(s - s_k)} \tag{7-10}$$

Design Equations

Given Ω_P, Ω_S, R_P, and A_S, two parameters are required to specify the Butterworth filter: the order N and the 3-dB cutoff frequency Ω_c. We want

$$-10 \cdot \log\left[\frac{1}{1 + \left(\frac{\Omega_P}{\Omega_c}\right)^{2N}}\right] \quad = \quad R_P$$

and

$$-10 \cdot \log\left[\frac{1}{1 + \left(\frac{\Omega_S}{\Omega_c}\right)^{2N}}\right] \quad = \quad A_S$$

Solving these two equations for N and Ω_c yields

$$N \quad = \quad \left\lceil \frac{\log[(10^{R_P/10} - 1)/(10^{A_S/10} - 1)]}{2 \cdot \log(\Omega_P/\Omega_S)} \right\rceil \tag{7-11}$$

and

$$\Omega_c = \frac{\Omega_P}{(10^{R_P/10} - 1)^{\frac{1}{2N}}} \qquad (7-12a)$$

or

$$\Omega_c = \frac{\Omega_S}{(10^{A_S/10} - 1)^{\frac{1}{2N}}} \qquad (7-12b)$$

The first value in the above equation satisfies specifications at Ω_P exactly while the second value satisfies specifications at Ω_S exactly. Now using equations (7-9), (7-10), (7-11), and (7-12), we can determine the system function $H_a(s)$ of a Butterworth filter.

Chebyshev Filter Design

There are two types of Chebyshev filters; one has equiripple behavior in the passband (Type-I) and the other has equiripple behavior in the stopband (Type-II). These two types are closely related and the design equations are essentially the same for both types. Therefore we will consider only Type-I filter called Chebyshev-I. The magnitude-squared response of an Nth-order Chebyshev-I filter is given by

$$|H_a(j\Omega)|^2 = \frac{1}{1 + \varepsilon^2 T_N^2\left(\frac{\Omega}{\Omega_c}\right)} \qquad (7-13)$$

where ε is a passband ripple factor related to K_1, and $T_N(\cdot)$ is an Nth order Chebyshev polynomial given by

$$T_N\left(\frac{\Omega}{\Omega_c}\right) = \begin{cases} \cos\left(N\cos^{-1}\left(\frac{\Omega}{\Omega_c}\right)\right), & 0 \le \Omega \le \Omega_c \\ \cosh\left(N\cosh^{-1}\left(\frac{\Omega}{\Omega_c}\right)\right), & \Omega_c \le \Omega \le \infty \end{cases}$$

The equiripple behavior of the Chebyshev filters is due to this polynomial.

A causal and stable $H_a(s)$ can be obtained by determining poles of $H_a(s)H_a(-s)$ and selecting the left-half poles for $H_a(s)$. The poles are obtained by finding the roots of

$$1 + \varepsilon^2 T_N^2\left(\frac{s}{j\Omega_c}\right) = 0$$

It can be shown [8] that, poles fall on an ellipse in the s-plane with the minor axis $\alpha\Omega$ along the real axis and major axis $\beta\Omega$ along the imaginary axis, where

$$\alpha = \frac{1}{2}\left(\gamma^{\frac{1}{N}} - \gamma^{-\frac{1}{N}}\right) \qquad (7-14a)$$

$$\beta = \frac{1}{2}\left(\gamma^{\frac{1}{N}} + \gamma^{-\frac{1}{N}}\right) \qquad (7-14b)$$

$$\gamma \;=\; \frac{1}{\varepsilon} + \sqrt{1 + \frac{1}{\varepsilon^2}} \tag{7-14c}$$

If $s_k = \sigma_k + j\Omega_k$ represents a pole of $H_a(s)$, then

$$\sigma_k \;=\; (\alpha\Omega_c)\cos\left[\frac{\pi}{2} + \frac{(2k+1)\pi}{2N}\right] \qquad k = 0, 1, \ldots, N-1 \tag{7-15a}$$

$$\Omega_k \;=\; (\beta\Omega_c)\sin\left[\frac{\pi}{2} + \frac{(2k+1)\pi}{2N}\right] \qquad k = 0, 1, \ldots, N-1 \tag{7-15b}$$

Now the system function is given by

$$H_a(s) \;=\; \frac{K}{\displaystyle\prod_{LHP\,poles} (s - s_k)} \tag{7-16}$$

where K is a normalizing factor chosen to make

$$H_a(0) \;=\; \begin{cases} 1 & N \quad odd \\[2mm] \dfrac{1}{\sqrt{1+\varepsilon^2}} & N \quad even \end{cases} \tag{7-17}$$

Design Equations

Given $\Omega_P, \Omega_S, R_P,$ and A_S, three parameters are required to specify a Chebyshev-I filter: the order N, the 3-dB cutoff frequency Ω_c, and the ripple factor ε. Equations for these parameters are given without proof.

$$\varepsilon \;=\; \sqrt{10^{R_P/10} - 1} \tag{7-18a}$$

$$\Omega_c \;=\; \Omega_P \qquad and \qquad \Omega_r \;\underline{\Delta}\; \frac{\Omega_S}{\Omega_P} \tag{7-18b}$$

$$A \;\underline{\Delta}\; 10^{\left(\frac{A_S}{20}\right)} \tag{7-18c}$$

$$g \;\underline{\Delta}\; \sqrt{(A^2 - 1)/\varepsilon^2} \tag{7-18d}$$

$$N = \left\lceil \frac{\log[g + \sqrt{g^2 - 1}]}{\log\left[\Omega_r + \sqrt{\Omega_r^2 - 1}\right]} \right\rceil \tag{7-18e}$$

Now using equations (7-14), (7-15), (7-16), (7-17) and (7-18), system function $H_a(s)$ can be determined.

7.3.2 Filter Transformations

The next step in IIR filter design is to transform the analog filters into digital filters. There are several complex-valued transformations that are available in the literature. The most popular among these are the *Impulse Invariant Transformation* and the *Bilinear Transformation*. The impulse invariant mapping preserves the impulse response of the analog filter by specifying the digital filter's impulse response to be a sampled version of the continuous-time response. As a result, the frequency response of the digital filter is an aliased version of the frequency response of the analog filter. The bilinear transformation is a one-to-one mapping which eliminates the aliasing problem by translating the analog frequency response to a digital system function with a comparable frequency response. Therefore in this section, we will briefly discuss the bilinear transformation.

It is defined by the following mapping from s to z.

$$s = \frac{2}{T}\frac{1-z^{-1}}{1+z^{-1}} \tag{7-19}$$

where T is a parameter. The inverse mapping is given by

$$z = \frac{1+\left(\frac{T}{2}\right)s}{1-\left(\frac{T}{2}\right)s} \tag{7-20}$$

To relate the frequency responses, let $z = e^{j\omega}$ and $s = j\Omega$ in equation (7-20). Then

$$e^{j\omega} = \frac{1+j\frac{\Omega T}{2}}{1-j\frac{\Omega T}{2}}$$

Solving for ω as a function of Ω, we obtain

$$\omega = 2\tan^{-1}\left(\frac{\Omega T}{2}\right)$$

The above equation gives a nonlinear compression effect between the analog frequency ω and the digital frequency ω. This is called *warping* of the analog frequencies, which also means that there is no aliasing. The entire imaginary axis in the s-plane is mapped uniquely onto the unit circle in the z-plane. This warping, however, is not a problem because digital frequencies can be *pre-warped* to obtain analog frequencies prior to designing analog filters. This pre-warping is given by

$$\Omega = \frac{2}{T}\tan\left(\frac{\omega}{2}\right) \tag{7-21}$$

Design Procedure

The design procedure to obtain an IIR lowpass digital filter can be summarized in the following steps. Given the design specifications:

- Passband cutoff frequency ω_P with ripple R_P
- Stopband cutoff frequency ω_S with attenuation A_S

1. Choose the parameter T and pre-warp the passband and stopband cutoff frequencies using equation (7-21) to obtain the corresponding Ω_P and Ω_S. This parameter T can be arbitrary and we will set it to 1.
2. Design an analog prototype filter $H_a(s)$ using the design equations in Section 7.3.1 for either the Butterworth or Chebyshev-I filter.
3. Set

$$H(z) \;=\; H_a\!\left(2\frac{1-z^{-1}}{1+z^{-1}}\right)$$

and simplify to obtain a rational $H(z)$ system function. This representation gives the direct form structures.
4. If desired, factor $H(z)$ into a product of biquad sections for the cascade form structure.

7.3.3 Spectral Transformations

In the last two sections, we described design techniques for digital lowpass filters. Certainly we would like to design other types of frequency selective filters such as highpass, bandpass, bandstop, or even other lowpass filters. This is accomplished by transforming the frequency-band of a lowpass filter so that it behaves like another frequency-selective filter. These transformations on the complex variable z are very similar to the bilinear transformation and the corresponding design equations are algebraic.

Let $H_{LP}(Z)$ be the designed lowpass digital filter and let $H(z)$ be the desired digital filter. Note that we are using two frequency variables, Z and z, with H_{LP} and H respectively. A mapping of the form:

$$Z^{-1} \;=\; G(z^{-1})$$

transforms

$$H_{LP}(Z)\big|_{Z^{-1}=G(z^{-1})} \to H(z)$$

if it is a valid mapping with proper parameters. The general form of the function $G(\cdot)$ is of an allpass filter type given by

$$Z^{-1} \;=\; G(z^{-1}) \;=\; \pm\prod_{k=1}^{n}\frac{z^{-1}-a_k}{1-a_k^{-1}}$$

where $|a_k| < 1$ for stability. By appropriately choosing n and the corresponding a_k's, we can obtain a variety of spectral transformations. The most widely used transformations are shown in Table 7-1.

Type of Transformation	Transformation	Parameters $Z = e^{j\omega'}, z = e^{j\omega}$ ω'_c : cutoff frequency of $H_{LP}(Z)$
Lowpass	$Z^{-1} \rightarrow \dfrac{z^{-1} - a}{1 - az^{-1}}$	$\omega_c =$ cutoff frequency of new filter $a = \dfrac{\sin[(\omega'_c - \omega_c)/2]}{\sin[(\omega'_c + \omega_c)/2]}$
Highpass	$Z^{-1} \rightarrow -\dfrac{z^{-1} + a}{1 + az^{-1}}$	$\omega_c =$ cutoff frequency of new filter $a = -\dfrac{\cos[(\omega'_c + \omega_c)/2]}{\cos[(\omega'_c - \omega_c)/2]}$
Bandpass	$Z^{-1} \rightarrow -\dfrac{z^{-2} - a_1 z^{-1} + a_2}{a_2 z^{-2} - a_1 z^{-1} + 1}$	$\omega_l =$ lower cutoff frequency $\omega_u =$ upper cutoff frequency $a_1 = -2\alpha K/(K+1)$ $a_2 = (K-1)/(K+1)$ $\alpha = \dfrac{\cos[(\omega_u + \omega_l)/2]}{\cos[(\omega_u - \omega_l)/2]}$ $K = \cot\dfrac{\omega_u - \omega_l}{2}\tan\dfrac{\omega'_c}{2}$
Bandstop	$Z^{-1} \rightarrow \dfrac{z^{-2} - a_1 z^{-1} + a_2}{a_2 z^{-2} - a_1 z^{-1} + 1}$	$\omega_l =$ lower cutoff frequency $\omega_u =$ upper cutoff frequency $a_1 = -2\alpha/(K+1)$ $a_2 = (1-K)/(1+K)$ $\alpha = \dfrac{\cos[(\omega_u + \omega_l)/2]}{\cos[(\omega_u - \omega_l)/2]}$ $K = \tan\dfrac{\omega_u - \omega_l}{2}\tan\dfrac{\omega'_c}{2}$

Table 7-1: Spectral Transformations for Digital Filters

7.3.4 Examples

In the following examples, we will design lowpass filter specified in Example 6-1 using both Butterworth and Chebyshev-I prototypes. This will allow us to compare FIR and IIR filter designs in terms of their orders and frequency responses. We will also examine differences between the two prototype designs.

Example 7-1:

In the first example, we will design the lowpass filter specified in Example 6-1. The design specifications are:

$$
\begin{aligned}
\text{Sampling frequency:} &\quad 8 \text{ KHz} \\
\text{Passband:} &\quad 0 - 1 \text{ KHz} \\
\text{Stopband:} &\quad 1.4 - 4 \text{ KHz} \\
\text{Passband Ripple:} &\quad 1 \text{ DB} \\
\text{Stopband Attenuation:} &\quad 50 \text{ DB}
\end{aligned}
$$

We will use a Butterworth filter as the analog prototype and the bilinear transformation method for digital filter design. The digital cutoff frequencies are:

$$
\omega_P \;=\; 2\pi\left(\frac{1}{8}\right) \;=\; 0.25\pi, \qquad \omega_S \;=\; 2\Pi\left(\frac{1.4}{8}\right) \;=\; 0.35\pi
$$

Following the design procedure outlined in section 7.3.2, the design steps are:

1. Pre-warped the analog frequencies: Using equation (7-21) with $T = 1$ we obtain,

$$
\Omega_P \;=\; 2\tan\left(\frac{\omega_P}{2}\right) \;=\; .82843, \qquad \Omega_S \;=\; 2\tan\left(\frac{\omega_S}{2}\right) \;=\; 1.22560
$$

2. Analog filter design (Butterworth Prototype): Using the design equations from Section 7.3.1 for the Butterworth filter, we obtain
   - from equation (7-11), $N = 17$, and
   - from equation (7-12b), $\Omega_C \;=\; .8735558$ which satisfies the stopband specifications exactly but exceeds those in the passband.

   The poles of $H_a(s)$ can now be computed from equation (7-9). These poles are shown in Table 7-2.

```
               -8.06016E-02 ± j   8.69829E-01
               -2.39060E-01 ± j   8.40208E-01
               -3.89377E-01 ± j   7.81975E-01
               -5.26435E-01 ± j   6.97113E-01
               -6.45566E-01 ± j   5.88511E-01
               -7.42712E-01 ± j   4.59868E-01
               -8.14566E-01 ± j   3.15565E-01
               -8.58682E-01 ± j   1.60515E-01
               -8.73556E-01 + j   0.00000E+00
```

Table 7-2: Poles of Analog Prototype Filter in Example 7-1 ———————————

- The gain in equation (7-10) is $\Omega_C^N = 1.00448$

This completes the analog filter design.

3. Bilinear transformation: By setting

$$H(z) \;=\; H_a\!\left(2\frac{1-z^{-1}}{1+z^{-1}}\right)$$

we obtain a rational $H(z)$ function whose numerator and denominator coefficients are shown in Table 7-3.

Filter Order : 17	Gain Factor : 7.41033E-09
Numerator Coefficients (Highest Order (in z) First)	Denominator Coefficients (Highest Order (in z) First)
1.000000E+00	1.000000E+00
1.700000E+01	-8.219959E+00
1.360000E+02	3.325991E+01
6.800000E+02	-8.7215E0E+01
2.380000E+03	1.650530E+02
6.188000E+03	-2.382394E+02
1.237600E+04	2.707250E+02
1.944700E+04	-2.467582E+02
2.430900E+04	1.822617E+02
2.430900E+04	-1.095289E+02
1.944700E+04	5.347435E+01
1.237600E+04	-2.106390E+01
6.188000E+03	6.605379E+00
2.380000E+03	-1.612977E+00
6.800000E+02	0.295961E+00
1.360000E+02	-3.846800E-02
1.700000E+01	3.145218E-03
1.000000E+00	-1.373291E-04

Table 7-3: Direct Form System Function in Example 7-1

4. Cascade form structure: Factoring $H(z)$ into a product of second-order sections we obtain the biquad coefficients shown in Table 7-4.

Filter Order : 17		Gain Factor : 7.41033E-09	
B_{1k}	B_{2k}	A_{1k}	A_{2k}
2.0000	1.0000	-1.272990	0.8732049
2.0000	1.0000	-1.131914	0.6656117
2.0000	1.0000	-1.024236	0.5071641
2.0000	1.0000	-0.942488	0.3868720
2.0000	1.0000	-0.881346	0.2969001
2.0000	1.0000	-0.837061	0.2317379
2.0000	1.0000	-0.807070	0.1876033
2.0000	1.0000	-0.789700	0.1620010
0.0000	1.0000	1.000000	-0.392002

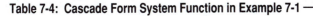

Table 7-4: Cascade Form System Function in Example 7-1

This completes the specified filter design in which the filter coefficients are obtained for both the direct and cascade form realization. Magnitude and log-magnitude plots of this filter are shown in Figures 7-7 and 7-8 respectively. Figure 7-9 shows the phase response while Figure 7-10 depicts the pole-zero pattern.

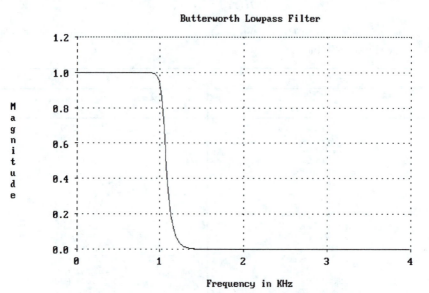

Figure 7-7: Butterworth Lowpass Filter: Magnitude Response

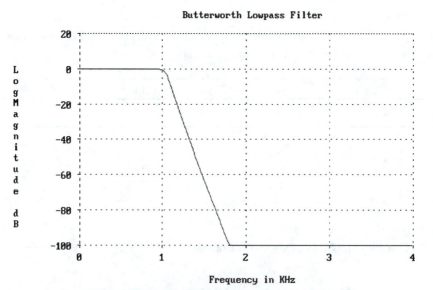

Figure 7-8: Butterworth Lowpass Filter: Log-Magnitude Response

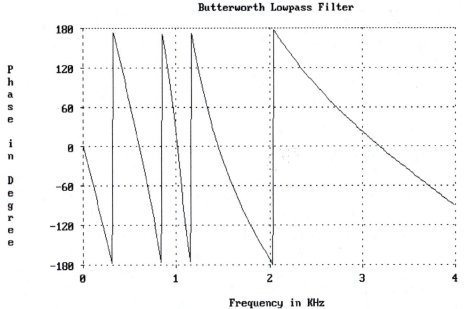

Figure 7-9: Butterworth Lowpass Filter: Phase Response

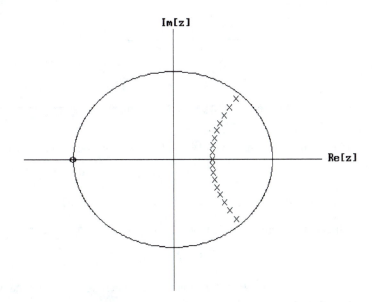

Figure 7-10: Butterworth Lowpass Filter: Magnitude Response

Example 7-2:

In this example we will design the same lowpass digital filter as in Example 7-1, by using a Chebyshev-I prototype. The design specifications are given in Example 7-1. We will also use the bilinear transformation. The digital cutoff frequencies are same as before:

$$\omega_p = 0.25\pi, \qquad \omega_s = 0.35\pi$$

The design steps are:

1. Pre-warped the analog frequencies:

$$\Omega_p = 2\tan\left(\frac{\omega_p}{2}\right) = .82843, \qquad \Omega_s = 2\tan\left(\frac{\omega_s}{2}\right) = 1.22560$$

2. Analog filter design (Chebyshev-I Prototype): Using the design equations from Section 7.3.1 for the Chebyshev-I filter, we obtain
    - from equation (7-18a), $\varepsilon = 0.50885$,
    - from equation (7-18b), $\Omega_c = .82843$ and $\Omega_r = 1.47942$,
    - from equation (7-18c), $A = 10^{\left(\frac{A_s}{20}\right)} = 316.22776$,
    - from equation (7-18d), $g = \sqrt{(A^2-1)/\varepsilon^2} = 621.45265$, and
    - from equation (7-18e), $N = 8$.

    The poles of $H_a(s)$ can now be computed from equations (7-14) and (7-15). These poles are shown in Table 7-5.

```
-2.90018E-02 ±  8.25487E-01
-8.25901E-02 ±  6.99814E-01
-1.23605E-02 ±  4.67601E-01
-1.45802E-02 ±  1.64200E-01
```

Table 7-5: Poles of Analog Prototype Filter in Example 7-2 ————————————

- The gain K in equation (7-16) is obtained from equation (7-17) as

$$K = \frac{1}{\sqrt{1+\varepsilon^2}} \prod_{LHP\,poles} |s_k|^2 = 0.03162$$

This completes the analog filter design.

3. Bilinear transformation: Setting

$$H(z) = H_a\left(2\frac{1-z^{-1}}{1+z^{-1}}\right)$$

we obtain a rational $H(z)$ function whose numerator and denominator coefficients are shown in Table 7-6.

4. Cascade form structure: Factoring $H(z)$ into a product of second-order sections we obtain the biquad coefficients shown in Table 7-7.

Filter Order : 8	Gain Factor : 6.71517E-06
Numerator Coefficients (Highest Order (in z) First)	Denominator Coefficients (Highest Order (in z) First)
1.000000E+00	1.000000E+00
8.000000E+00	-6.065304E+00
2.800000E+01	1.719893E+01
5.600000E+01	-2.953864E+01
7.000000E+01	3.345596E+01
5.600000E+01	-2.552769E+01
2.800000E+01	1.230249E+01
8.000000E+00	-3.858937E+00
1.000000E+00	0.535854E+00

Table 7-6: Direct Form System Function in Example 7-2 ——————————

Filter Order : 8		Gain Factor : 6.715176E-06	
B_{1k}	B_{2k}	A_{1k}	A_{2k}
2.0000	1.0000	-1.3829	0.9516
2.0000	1.0000	-1.4516	0.8631
2.0000	1.0000	-1.5930	0.7909
2.0000	1.0000	-1.7065	0.7482

Table 7-7: Cascade Form System Function in Example 7-2 ——————————

This completes the specified filter design in which the filter coefficients are obtained for both the direct and cascade form realizations. Magnitude and log-magnitude plots of this filter are shown in Figures 7-11 and 7-12 respectively. Figure 7-13 shows the phase response while Figure 7-14 depicts the pole-zero pattern.

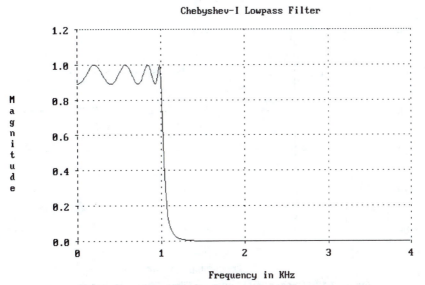

Figure 7-11: Chebyshev-I Lowpass Filter: Magnitude Response

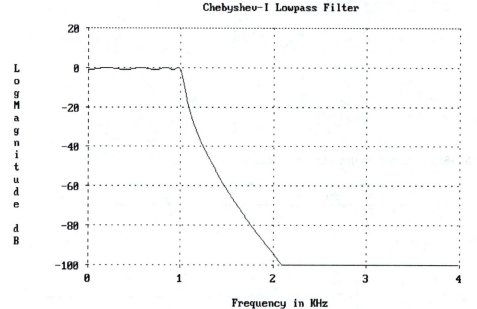

Figure 7-12: Chebyshev-I Lowpass Filter: Log-Magnitude Response

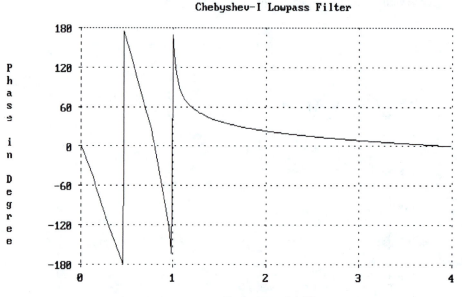

Figure 7-13: Chebyshev-I Lowpass Filter: Phase Response

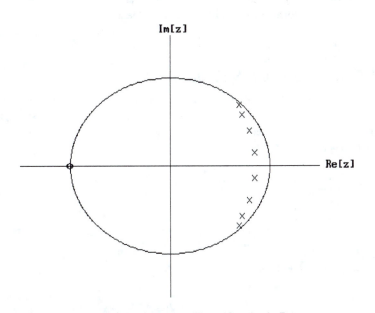

Figure 7-14: Chebyshev-I Lowpass Filter: Magnitude Response

Comparing these examples with those in Chapter 6, we make several observations. IIR filters generally have much lower order and sharper transition band than FIR filters, which make them attractive from an implementation point of view. However, IIR filters exhibit a nonlinear phase response. IIR filter designs also vary markedly from the analog prototypes used. For a given set of specifications, Butterworth filters have a higher order than Chebyshev filters, while Elliptic filters are the optimum in this respect. From the phase plots, it is clear that the Chebyshev filter has more nonlinear response than the Butterworth filter, while the elliptic filter has the most nonlinear phase response.

Even though it is possible to design other frequency selective filters using the methods outlined in this chapter, several excellent computer programs are currently available for use on personal computers to design such filters. Therefore, we will now focus on implementing designed filters on the ADSP-2101 microcomputer.

7.4 DIRECT FORM IIR FILTER

As discussed in section 7.2, realization of an IIR filter can take many forms. The most useful of those in the ADSP-2101 implementation are the Direct Form and Cascade Form structures. In this section, we describe a subroutine for the single precision direct form structure. This form is easier to implement than the cascade form because the scaling mechanism to avoid overflow and increase dynamic range is simpler to understand. Similarly the effect of limit-cycle oscillations is well understood. In the cascade form, each biquad section is implemented using the direct form. Therefore it is necessary to study the direct form subroutine.

Of the two possible implementations of the direct form structure, the Direct Form II is the most widely used form since it contains only one delay line. We begin with the description of a direct form II subroutine followed by an example program. The emphasis in the example will be on understanding the scaling operation.

7.4.1 A Direct Form II Subroutine

The direct form II structure is described by equation (7-1) (or (7-2)) and is shown in Figure 7-2. It has two parts: a feedforward (or FIR like) part and a feedback (or recursive) part. It can best be analyzed using an intermediate signal $w(n)$ shown in Figure 7-2. In terms of this signal, equation (7-1) can be written as

$$H(z) = \frac{Y(z)}{W(z)} \cdot \frac{W(z)}{X(z)}$$

where

$$\frac{Y(z)}{W(z)} = b_0 + b_1 z^{-1} + \cdots + b_N z^{-N}$$

is the feedforward section, and

$$\frac{W(z)}{X(z)} = \frac{1}{1 - a_1 z^{-1} - \cdots - a_N z^{-N}}$$

is the feedback section. The corresponding difference equations are

$$y(n) = b_0 w(n) + a_1 w(n-1) + \ldots + b_N w(n-N) \qquad (7-22)$$

and

$$w(n) = x(n) + a_1 w(n-1) + \ldots + a_N w(n-N) \qquad (7-23)$$

Therefore the only signal that needs to be stored is $w(n)$ which requires a single circular buffer. The subroutine that implements these two equations is shown in Listing (7-1). Before the execution of the subroutine, the input value, $x(n)$, is in register MR1 and the signal $w(n)$ values along the delay line are shifted by one sample in the circular buffer. In the subroutine, the sum-of-products of a values and the delay line values are computed first according to equation (7-22). This replaces the previous value of $w(n)$ by the current update. Next, the sum-of-products of the b values and the delay line values, with $w(n)$ updated, are computed according to equation (7-23) and the result stored in MR1 which, at the conclusion of the subroutine, provides the output value $y(n)$. Finally, the circular buffer index register I0 is modified so that the delay line current location is $w(n-1)$.

The above subroutine requires a total of $2N + 11$ cycles for a filter of order N. At an 8KHz sampling rate and an instruction cycle time of 80 nanoseconds, this permits a filter of length 700 with 150 instruction cycles for other operations.

```
{ IIR Direct Form II Filter Subroutine                      IIRDFMII.DSP

    Calling Parameters
          MR1 = Input sample ( x[n] )
          MR0 = 0
          I0 --> Delay line buffer current location ( w[n-1] )
          L0 = Filter length
          I5 --> Feedback coefficients (a's)
          L5 = Filter length - 1
          I6 --> Feedforward coefficients (b's)
          L6 = Filter length
          M0 = 0
          M1,M4 = 1
          M2 = 2
          CNTR = Filter length - 2
          AX0 = Filter length - 1

    Return Values
          MR1 = output sample ( y[n] )
          I0 --> delay line current location ( w[n-1] )
          I5 --> feedback coefficients
          I6 --> feedforward coefficients

    Altered Registers
          MX0,MY0,MR

    Computation Time
          (N - 2) + (N - 1) + 10 + 4 cycles  (N = M = Filter order)

    All coefficients and data values are assumed to be in 1.15 format.
}

.MODULE iirdfmii_sub;
.ENTRY  iir_dfii;

iir_dfii: MX0=DM(I0,M1), MY0=PM(I5,M4);
          DO poleloop UNTIL CE;
poleloop: MR=MR+MX0*MY0(SS), MX0=DM(I0,M1), MY0=PM(I5,M4);
          MR=MR+MX0*MY0(RND);
          IF MV SAT MR;
          CNTR=AX0;
          DM(I0,M0)=MR1;                              { Update w(n) }
          MR=0, MX0=DM(I0,M1), MY0=PM(I6,M4);
          DO zeroloop UNTIL CE;
zeroloop: MR=MR+MX0*MY0(SS), MX0=DM(I0,M1), MY0=PM(I6,M4);
          MR=MR+MX0*MY0(RND);
          MODIFY (I0,M2);
          RTS;
.ENDMOD;
```
——————— **Listing 7-1: IIR Direct Form II Subroutine** ———————————

7.4.2 An Example Program

As an example of the above subroutine, we provide a complete example program to implement the lowpass filter designed in Example 7-1. This program is shown in Listing 7-2. The filter coefficients are given in Table 7-3 and the magnitude response is shown Figure 7-7.

The most important issue in IIR filter implementation is the proper scaling of the coefficients to avoid overflow at the MAC output after every sum-of-products calculations. On the other hand, excessive scaling to avoid overflow may result in inefficient use of the available dynamic range. Therefore the filter coefficients must be carefully scaled. This problem is not an important one in the case of FIR filters because these filters are generally designed to have

```
{ Lowpass IIR Filter                                           IIRDF2LP.DSP
          Using Direct Form II Structure

    This program implements a single precision Direct Form II IIR filter
    structure using subroutine iir_dfii available in disk file IIRDFMII.DSP.
    The filter coefficients used in this program are for the lowpass filter
    described in Example 7-1.  These coefficients are stored in disk files:
    IIRDFLPA.DAT and IIRDFLPB.DAT.

    This program is written for EZ-ICE and EZ-LAB system with EZLAB1.ACH
    architecture file.  Additionally, this program uses subroutine
    CntlReg_inits to initialize all control registers.  It is available in
    disk file CREGINIT.DSP.  Assemble IIRDF2LP.DSP, IIRDFMII.DSP, and
    CREGINIT.DSP using ASM21.EXE.  Link using LD21.EXE to produce
    IIRDF2LP.EXE.  Load IIRDF2LP.EXE in EZ-ICE and execute.
}

.MODULE/RAM/ABS=0/BOOT=0              IIR_DF2LP;
.PORT             write_dac0;
.PORT             load_dac;
.CONST            length=18; {N = 17}
.VAR/CIRC         data[length];
.VAR/PM/CIRC      a_coeffs[length];
.VAR/PM/CIRC      b_coeffs[length];
.VAR/DM/CIRC      delay_line[length];
.INIT             a_coeffs: <IIRDFLPA.DAT>;
.INIT             b_coeffs: <IIRDFLPB.DAT>;
{------------------------------------------------------------------------}
     JUMP START; RTI; RTI; RTI;        {Reset Vector}
     RTI; RTI; RTI; RTI;               {irq2}
     RTI; RTI; RTI; RTI;               {sport0 TX}
     jump sample; RTI; RTI; RTI;       {sport0 RX}
     RTI; RTI; RTI; RTI;               {irq0}
     RTI; RTI; RTI; RTI;               {irq1}
     RTI; RTI; RTI; RTI;               {timer}
{--- Initialize ---------------------------------------------------------}
start:  CALL CntlReg_inits;            { set up SPORTS, TIMER, etc }
        I0=^delay_line; M0=0; L0=length;
        I5=^a_coeffs;    M4=1; L4=length;
        I6=^b_coeffs;    M5=1; L6=length;
                         m1=1;
                         m2=2;
        AX0=length-1;
        CNTR=length;
        DO zero UNTIL CE;
zero:   DM(I0,M1)=0;            { clear out the filter delay line buffer }
        CNTR=length-2;
        ICNTL=B#00111;         { disable IRQ nesting, all IRQs edge-sensitive}
        IMASK=B#101000;        { enable IRQ2 and SPORT0_RX interrupt }
wait:   IDLE;
        JUMP wait;
{--- Process Input Sample ----------------------------------------------}
sample: MR1=RX0;               { get new sample from SPORT0 (from microphone)}
        CALL iir_dfii;
        TX0=MR1;                         { filtered output to SPORT (to spkr) }
        SR = ASHIFT MR1 BY 1 (HI);
        DM(write_dac0)=SR1;            { latch sample for D/A }
        DM(load_dac)=SR1;              { display sample on oscilloscope }
        RTI;
.ENDMOD;
```

—————————— **Listing 7-2: Example Program using Direct Form II Structure** ——————————

maximum gain equal to one at any frequency. Although the frequency domain unity gain condition does not absolutely preclude overflow, it can make overflow acceptably improbable while using the available dynamic range more efficiently. Furthermore, the output of an FIR filter is never fed back, thus any error due to overflow does not affect subsequent outputs. An IIR filter is also designed to have maximum unity gain at any frequency. However, it has feedforward and feedback parts and each part is implemented separately as a sum-of-products calculation. Now there is no guarantee that each part will have a maximum gain equal to one. Therefore direct use of the filter coefficients from the design (e.g. such as those from Table 7-3) can result in severe overflow problems. Moreover, because some of the values are fed back into the filter calculations, the error can corrupt the subsequent output values.

Referring to equations (7-22) and (7-23), it is obvious that we have to protect the intermediate signal $w(n)$ from overflowing at each n. The ADSP-2101 processor, in its default mode, implements 1.15 fixed-point format arithmetic. Therefore the maximum value of data is 1. Equation (7-22) is an FIR filter and hence its gain is very easy to calculate. Consider

$$
|y(n)| = \left| \sum_{k=0}^{N} b_k w(n-k) \right|
$$

$$
\leq \sum_{k=0}^{N} |b_k| \, |w(n-k)| \leq \sum_{k=0}^{N} |b_k|
$$

since $|w(n-k)| \leq 1$. Since the output values also satisfy $|y(n)| \leq 1$, we must have

$$
\sum_{k=0}^{N} |b_k| \leq 1
$$

From Table 7-3, all b_k values are positive and their sum is equal to 131,068. Hence these coefficients must be scaled by 7.62962×10^{-6} to avoid overflow in the feedforward path. These scaled coefficients in 1.15 hexadecimal format are available in disk file IIRDFLPB.DAT. An alternate approach is to scale the b_k values so that

$$
|B(e^{i\omega})|_{\text{max over}\,\omega} = 1
$$

where $B(e^{j\omega})$ is the discrete-time Fourier transform of b_k.

The gain for the feedback part is not so easy to calculate. We can compute the impulse response of the feedback section and then use the above approach. However this will be tedious and in fact may lead to overscaling. The easier method is to use the alternate approach in frequency domain. From equation (7-23), the magnitude response of the feedback section is

$$
\left| \frac{W(e^{j\omega})}{X(e^{j\omega})} \right| = \frac{1}{\left| 1 + \sum_{k=1}^{N} a_k e^{-j\omega} \right|}
$$

This response can be computed using a computer program. The a_k coefficients can now be scaled so that the maximum value of the above response is 1. For the values given in Table 7-3, this maximum gain is 9.01303×10^2. Hence these coefficients must be scaled by 1.1095×10^{-3}. Another approach would be to look at the overall gain factor (7.41033×10^{-9}) given in Table 7-3. Since we know the gain for the FIR part (7.62962×10^{-6}), the gain for the feedback part is 0.97125×10^{-3}, which agrees with the earlier value. The scaled a_k coefficients in 1.15 hexadecimal format are available in the disk file IIRDFLPA.DAT.

7.4.3 Limit Cycles in Direct Form Structures

The overflow problems described above, if not properly treated, can cause severe problems known as *large-scale limit cycles*. If a large value results in a sum-of-products calculation of the intermediate signal $w(n)$, then in a two's-complement arithmetic, what should have been a large positive number may instead become a large negative number. This large number is fed back causing another large number with sign reversal. Once these large oscillations start they are difficult to stop without turning off the hardware. However, a proper scaling discussed in the above section can virtually eliminate these large-scale limit cycles. Moreover, an instruction to saturate the result after overflow detection can prevent sign reversal.

There is another type of limit cycle which is completely different from the one above. It results from the quantization effects in calculations and is called *small-scale limit cycles*. When the input signal is zero, every stable filter's output decays to zero over time. When this output is smaller than the quantization step, it may get rounded up to the previous level causing steady oscillations with small amplitudes. When there is a significant amount of input, these limit cycles tend to go away or at least are not noticeable. This is not a severe problem however, but in audio applications, it can result in an irritating tone when there is no input.

7.4.4 Exercises

1. Write a subroutine to implement the Direct Form I structure. This will require two circular buffers, one each for input and output. Assume that the numerator and denominator orders of $H(z)$ are different. Use properly scaled filter coefficients from example 7-1 and verify its operation.

2. Implement the lowpass Chebyshev-I filter designed in Example 7-2 in the Direct Form I structure. Properly scale coefficients using ideas developed in Section 7.4.2. Verify its operation and compare this filter with the Butterworth filter.

3. Limit cycles can easily be simulated by causing overflow to occur. Modify the IIRDFMII.DSP subroutine by eliminating the saturation logic. Now try to cause an overflow either by underscaling the a_k coefficients or increasing the level of the input signal. You should be able obtain an annoyingly large tone. This is the large-scale limit cycle. To simulate a small-scale limit cycle, turn off the input and observe the output signal. Since the output doesn't always enter into this limit cycle, you will have to try turning off the input several times to observe one.

7.5 CASCADE FORM IIR

The direct form structure suffers from many practical problems even though it executes faster. The coefficients and data must be scaled all at once, which gives rise to larger numerical errors. The poles of the single-stage direct form get increasingly sensitive to quantization errors. These problems can be alleviated if a cascade of second order biquad IIR filter sections are used in the implementation of IIR filters. The biquad implementation executes slower but generates smaller numerical errors. The biquads can be scaled separately and then cascaded in order to minimize the coefficient quantization and the recursive numerical errors.

7.5.1 A Cascade Form Subroutine

A second-order biquad IIR filter section is shown in Figure 7-3. Its system function in the z-domain is given by

$$H(z) \;=\; \frac{Y(z)}{X(z)} \;=\; \frac{B_0 + B_1 z^{-1} + B_2 z^{-2}}{1 - A_1 z^{-1} - A_2 z^{-2}}$$

The corresponding difference equation for a biquad section is

$$y(n) \;=\; B_0 x(n) + B_1 x(n-1) + B_2 x(n-2) + A_1 y(n-1) + A_2 y(n-2)$$

which can also be obtained from Figure 7-3.

An ADSP-2101 subroutine that implements a cascade structure is shown in Listing 7-3. The subroutine is arranged as a module and is labeled *biquad_sub*. There are a number of registers that need to be initialized in order to execute this subroutine. It may be sufficient to do this initialization only once (e.g., at powerup) if other executed algorithms do not need these registers. In most typical cases, however, some of these registers may need to be set every time the *biquad_sub* routine is called. It may sometimes be beneficial, from a modular software point of view, to always initialize all the setup registers as a part of this subroutine.

The *biquad_sub* routine takes its input from the SR1 register. This register must contain the 16-bit input $x(n)$. $x(n)$ is assumed to be already computed before this subroutine is called. The output of the filter is also made available in the SR1 register.

After the initial design of a high order IIR filter, all coefficients must be scaled down in each biquad stage separately. This is necessary in order to conform to the 16-bit fixed-point fractional number format as well as to ensure that overflows will not occur in the final multiply-accumulate operations in each stage. The scaled-down coefficients are the ones that get stored in the processor's memory. The operations in each biquad are performed with scaled data and coefficients are eventually scaled up before being output to the next one. The choice of a proper scaling factor depends greatly on the design objectives, and in some cases it may even be unnecessary.

During the initialization of the *biquad_sub* routine, the index register I0 points to the data memory buffer that contains the previous error inputs and the previous biquad section outputs. This buffer must be initialized to zero at powerup unless some nonzero initial condition is desired. The index register I1 points to another buffer in data memory that contains the individual scale factors for each biquad. The buffer length register L1 is set to zero if the filter has only one biquad section. In the case of multiple biquads, L1 is initialized with the number of biquad sections. The index register I4, on the other hand, points to the circular program memory buffer that contains the scaled biquad coefficients. These coefficients are stored in the order: $B_2, B_1, B_0, A_2,$ and A_1 for each biquad. All of the individual biquad coefficient groups must be stored in the same order that the biquads are cascaded in. The buffer length register L4 must be set to the value of ($5 \times$ number of biquad sections). Finally, the loop counter register CNTR must be set to the number biquad sections since the filter code will be executed as a loop.

The core of the *biquad_sub* routine starts its execution at the *biquad* label. The routine is organized in a looped fashion where the end of the loop is the instruction labeled *sections*. Each iterations of the loop executes the computations for one biquad. The number of loops to be executed is determined by the CNTR register contents. The SE register is loaded with the appropriate scaling factor for the particular biquad at the beginning of each loop iterations. After this operation, the coefficients and the data values are fetched from memory in the sequence that they have been stored. These numbers are multiplied and accumulated until all of the values for a particular biquad have been accessed. The result of the last multiply/accumulate is rounded to 16 bits and upshifted by the scaling value. At this point the *biquad* loop is executed again, or the filter computations are completed by doing the final update to the delay line. The delay lines for data values are always being updated within the *biquad* loop as well as outside of it.

The filter coefficients must be scaled appropriately so that overflows occur after the upshifting operation between the biquads. If this is not ensured by design, it may be necessary to include some overflow checking between the biquads.

```
{ Cascaded Biquad IIR Filter Subroutine                        IIRCASFM.DSP

      Calling Parameters
            SR1 = input sample
            I0 --> delay line buffer for x(n-2), x(n-1), y(n-2), y(n-1)
            L0 = 0
            I1 --> list of scale factors for each biquad section
            L1 = 0 (in the case of a single biquad)
            L1 = number of biquad sections
            I4 --> scaled coefficients b2,b1,b0,a2,a1,  b2,b1,b0,a2,a1, ...
            L4 = 2.5 * filter order --or-- 5 * number of biquad sections
            M0,M4 = 1
            M1 = -3
            M2 = 1   (in the case of multiple biquads)
            M2 = 0   (in the case of a single biquad)
            M3 = (1 - length of delay line buffer)
            CNTR = number of biquad sections

      Return Values
            SR1 = output sample
```

```
              I0 --> inside delay line buffer
              I1 --> top of scale factor list
              I4 --> top of coefficients

        Altered Registers
              MX0,MX1,MY0,MR,SE,SR

        Computation Time
              8 * number of biquad sections + 5 + 5 cycles
        All coefficients and data values are assumed to be in 1.15 format.
}

.MODULE biquad_sub;
.ENTRY  biquad;
biquad: I0=^delayline;
        DO sections UNTIL CE;
          SE=DM(I1,M2);
          MX0=DM(I0,M0), MY0=PM(I4,M4);                      {get x(n-2), B2}
          MR=MX0*MY0(SS), MX1=DM(I0,M0), MY0=PM(I4,M4);      {get x(n-1), B1}
          MR=MR+MX1*MY0(SS), MY0=PM(I4,M4);                  {get         B0}
          MR=MR+SR1*MY0(SS), MX0=DM(I0,M0), MY0=PM(I4,M4);   {get y(n-2), A2}
          MR=MR+MX0*MY0(SS), MX0=DM(I0,M1), MY0=PM(I4,M4);   {get y(n-1), A1}
          DM(I0,M0)=MX1, MR=MR+MX0*MY0(RND);   {store x(n-1) as new x(n-2)}
sections: DM(I0,M0)=SR1, SR=ASHIFT MR1 (HI);       {store x(n) as new x(n-1)}
        DM(I0,M0)=MX0;
        DM(I0,M3)=SR1;
        RTS;

.ENDMOD;
```
——————————— **Listing 7-3: IIR Cascade Form Subroutine** ———————————

7.5.2 An Example Program

As an example of the above subroutine, we once again consider the lowpass filter designed in Example 7-1. The appropriate coefficients for this program are the cascade form coefficients shown in Table 7-4. Note that B_0 coefficients for all biquad sections are equal to 1 in our formulation according to equation (7-3).

The scaling issue in the cascade form is even more trickier than the direct form. As discussed earlier, each biquad section must be scaled appropriately to avoid overflow of intermediate signals (the total number of which is equal to the number of biquads). Each biquad is in fact a direct form IIR filter. Therefore the same analysis of scaling described in section 7.4.2 must be performed for each section. The FIR part is relatively easy and since each B_k is positive and their sum is equal to 4, each B_k must be scaled by 0.25 to avoid overflow in the FIR part. The feedback part of each biquad section must be analyzed individually and its maximum frequency domain gain must be computed. The inverse of this gain now forms the scaling factor for that section. These scaling factors are stored in disk file IIRCFLPS.DAT while the scaled coefficients are available the IIRCFLPC.DAT file.

```
{ Lowpass IIR Filter                                         IIRCFMLP.DSP
        Using Cascade Form Structure

   This program implements a single precision Cascade Form IIR filter
   structure using subroutine biquad_sub available in disk file
   IIRCASFM.DSP. The filter coefficients used in this program are for the
   lowpass filter described in Example 7-1.  These coefficients are stored
   in disk file IIRCFLPC.DAT and the scaling factors in IIRCFLPS.DAT.
```

```
    This program is written for EZ-ICE and EZ-LAB system with EZLAB1.ACH
    architecture file.  Additionally, this program uses subroutine
    CntlReg_inits to initialize all control registers.  It is available in
    disk file CREGINIT.DSP.  Assemble IIRCFMLP.DSP, IIRCASFM.DSP, and
    CREGINIT.DSP using ASM21.EXE.  Link using LD21.EXE to produce
    IIRCFMLP.EXE.  Load IIRCFMLP.EXE in EZ-ICE and execute.
}

.MODULE/RAM/ABS=0/BOOT=0                      IIR_DF2LP;
.PORT              write_dac0;
.PORT              load_dac;
.CONST  N = 9;                   { number of biquad sections, example: 9 }
.CONST  N_x_5 = 45;              { number of biquad sections times five }
.VAR/DM delayline[4];            { this is scratchpad memory }
.VAR/DM scalelist[N];        { initialize with the scale factor for each biquad
}
.VAR/PM coefflist[N_x_5];    { init with filter coefficients for each biquad
}
.INIT             scalelist: <IIRCFLPS.DAT>;
.INIT             coefflist: <IIRCFLPC.DAT>;
{-----------------------------------------------------------------------}
    JUMP START; RTI; RTI; RTI;       {Reset Vector}
    RTI; RTI; RTI; RTI;              {irq2}
    RTI; RTI; RTI; RTI;              {sport0 TX}
    jump sample; RTI; RTI; RTI;      {sport0 RX}
    RTI; RTI; RTI; RTI;              {irq0}
    RTI; RTI; RTI; RTI;              {irq1}
    RTI; RTI; RTI; RTI;              {timer}
{--- Initialize ------------------------------------------------------}
start:  CALL CntlReg_inits;          { set up SPORTS, TIMER, etc }
        I0=^delayline;  M0=1;  L0=0;
        I1=^scalelist;  M1=-3; L1=N;
        I4=^coefflist;  M4=1;  L4=N_x_5;
        I6=^b_coeffs;   M5=1;  L6=length;
                        M2=1;
                        M3=-3;

        CNTR=4;
        DO zero UNTIL CE;
zero:   DM(I0,M1)=0;                 { clear out the filter delay line buffer }
        CNTR=N;
        ICNTL=B#00111;               { disable IRQ nesting, all IRQs edge-sensitive}
        IMASK=B#101000;              { enable IRQ2 and SPORT0_RX interrupt }
wait:   IDLE;
        JUMP wait;
{--- Process Input Sample -------------------------------------------}
sample: SR1=RX0;                     { get new sample from SPORT0 (from microphone)}
        CALL biquad;
        TX0=SR1;                     { filtered output to SPORT (to spkr) }
        SR = ASHIFT SR1 BY 1 (HI);
        DM(write_dac0)=SR1;          { latch sample for D/A }
        DM(load_dac)=SR1;            { display sample on oscilloscope }
        RTI;

.ENDMOD;
```

————————— **Listing 7-4: Example Program using Cascade Form Structure** —————————

7.5.3 Limit Cycles in Cascade Form Structures

In the case of the Cascade Form structure with more than one biquad, the limit cycles are much more difficult to analyze. When any biquad section, except the last one, exhibits a small-scale limit cycle, the output limit cycle is filtered by the succeeding sections. If the frequency of this limit cycle fall near the resonance frequency in a succeeding section, the amplitude of the limit cycle becomes very large. Likewise, if the succeeding section has a null at or near the limit cycle frequency then the limit cycle is virtually eliminated. Therefore the problem of small-scale limit cycles can be minimized by properly rearranging the biquad sections to filter out observed limit cycles.

7.5.4 Exercises

1. Implement the lowpass Chebyshev-I filter designed in Example 7-2 in a Cascade Form structure. Use Table 7-7 and properly scale numerator coefficients. Determine scaling factor for each section using the ideas developed above 7.5.2. Verify its operation and compare this filter with the Butterworth filter of Exercise 1 in Section 7.4.5.

2. Design a digital filter with the following specifications:

    > Sampling frequency : 8 KHz
    > Passband : 0 – 2 KHz
    > Passband ripple : 1 dB
    > Prototype : Butterworth
    > Filter order: 10
    > Filter Transformation : Bilinear

    Determine the coefficients for the Cascade Form realization. Properly scale these coefficients and implement the filter using the IIRCASFM subroutine. Verify the operation of the filter.

7.6 ALL-POLE LATTICE FILTER

Section 6.5 described all-zero lattice filter implementation. From the discussion in Section 7.2.3, it is obvious that all-pole lattice is very similar to the all-zero lattice except for the roles of input and output. Therefore the implementation of the all-pole lattice filter on the ADSP-2101 is also similar to that of the all-zero lattice filter. This implementation is shown in Listing 7-5.

Before the *all_pole_lattice_filter* routine is called, various registers must be preloaded. The index register I0 should point to the start of the input buffer, I1 to the start of the coefficient buffer, I2 to the start of the output buffer, and I4 to the start of the filter delay line. The length registers L0 and L2 should both be set to zero, and L1 and L4 should be set to the filter order to make the coefficient and the delay line buffers circular. The modify registers M0 and M4 should both be set to one; M1 and M5 should both be set to -1. The M6 should be set to three and M7 to -2. The SE register, which controls data scaling, should be set to an appropriate value, and the AX0 should be set to the order of the filter less one.

The routine loads the first input data value into the MY0. The *outloop* loop is executed one for each output data value. The MR register is loaded with the scaled value of $f_m(n)$ at the same time the coefficient k_m and delay line value $g_{m-1}(n-1)$ are loaded. The next instruction computes the value $f_{m-1}(n)$ and also loads the next multiplier operands. The *dataloop* loop performs the remainder of the filtering operation on the data point.

In the *dataloop* loop, $f_{m-1}(n)$ is computed and then shifted to the proper format for the next multiplication. Then the value of $g_m(n)$ is computed and stored in the delay line. After the *dataloop* loop has been executed, the pointers to the delay line and the coefficient buffer are moved to the tops of their buffers at the same time the output of the filter and the last delayed point $f_1(n)$ are saved.

```
{   All-Pole Lattice Filter Subroutine                          IIRLATFL.DSP

            Calling Parameters
                    CNTR = Length of Excitation Signal
                    I0 --> Excitation Signal                 L0 = 0
                    I1 --> Coefficient Buffer (circular)     L1 = Filter Order
                    I2 --> Output Buffer                     L2 = 0
                    I4 --> Delay Line Buffer (circular)      L4 = Filter Order
                    AR = Formatted 1
                    M0,M4 = 1                                M1,M5 = -1
                    M6 = 3                                   M7 = -2
                    SE = Appropriate scale value
                    AX0 = Filter Order - 1

            Return Values
                    Output Buffer Filled

            Altered Registers
                    MX0,MY0,MY1,MR,SR,I0,I1,I2,I4

            Computation Time
                    (6*(Filter Length-1)+8) * Output Buffer Length + 3 + 6 cycles
}

.MODULE all_pole_lattice_filter;
.ENTRY  p_latt;

p_latt: MY0=DM(I0,M0);                              {Get input data}
        DO outloop UNTIL CE;                        {Loop through output}
                CNTR=AX0;
                MR=AR*MY0 (SS), MX0=DM(I1,M0), MY0=PM(I4,M4);      {Get g,k}
                MR=MR-MX0*MY0 (SS), MX0=DM(I1,M0), MY0=PM(I4,M4); {MR=f1}
                DO dataloop UNTIL CE;               {Loop through filter}
                        MR=MR-MX0*MY0 (SS);         {Compute fm}
                        SR=ASHIFT MR1 (HI);         {Reformat fm}
                        MY1=SR1, MR=AR*MY0 (SS);    {Format gm+1}
                        MR=MR+MX0*MY1 (SS), MX0=DM(I1,M0),MY0=PM(I4,M7);
                                                                  {MR=gm}
                        SR=ASHIFT MR1 (HI);         {Reformat gm}
dataloop:               PM(I4,M6)=SR1, MR=AR*MY1 (SS);{Save gm format fm}
                MY0=PM(I4,M7), MX0=DM(I1,M1);       {Reset Pointers}
                MY0=DM(I0,M0), SR=ASHIFT MR1 (HI);  {Get new data point}
                DM(I2,M0)=MY1, SR=SR OR LSHIFT MR0 (LO); {Store output}
outloop:        PM(I4,M4)=SR1;                      {Save Y}
        RTS;
.ENDMOD;
```

Listing 7-5: All-Pole Lattice Filter Subroutine

7.7 SUMMARY

We have briefly discussed the design of frequency selective IIR digital filters using analog prototypes. Among all analog prototypes, elliptic filters are the most efficient to implement but we need commercially available filter design programs for carrying out the design. In practice, these filters should be considered when phase distortion is unimportant. Butterworth filters, on the the hand, give mild nonlinear phase responses.

The implementation of IIR filters on the ADSP-2101 was the most important topic of this chapter. We described the Direct Form and Cascade Form structures which are suitable for implementation on the ADSP-2101. We then discussed assembly language subroutines for these structures along with complete example programs. In each case we provided a thorough and useful discussion on coefficient scaling to avoid overflow and large-scale limit cycle problems. Finally, we presented the all-zero lattice filter structure and its implementation on the ADSP-2101 microcomputer.

chapter 8

FAST FOURIER TRANSFORM
IMPLEMENTATIONS

8.1 INTRODUCTION

Signal representation and analysis in the frequency domain is an important area in digital signal processing. It is performed using the discrete Fourier transform (DFT) on the data sequence. The DFT is widely used in applications to determine spectral contents, to perform correlation analysis and to implement linear filtering in the frequency domain. This widespread use is due to the existence of efficient algorithms for computing the DFT. In this chapter, we discuss the DFT and describe two common methods of efficient computation, called fast Fourier transforms (FFT), and explain their detailed implementation on the ADSP-2101.

The architecture and the instruction set of the ADSP-2101 processor is very well suited for implementation of the FFT algorithms. These algorithms repeatedly perform a core multiply/add operation, called a *butterfly*, on ordered pairs of data points and either require a scrambled order of the input data or result in a scrambled order of the output data. This requires a very efficient address generation which can be accomplished by two data address generators (DAG0 and DAG1) of the ADSP-2101. Furthermore, DAG1 contains a bit-reversal hardware which can be used to scramble or unscramble the data on the fly.

The chapter begins with a brief review of the DFT in Section 8.2. Two FFT algorithms are then considered in the next two sections. In Section 8.3, the decimation-in-time (DIT) FFT is first described and then a complete discussion on its implementation using program modules is provided. A similar treatment is done in Section 8.4 for the decimation-in-frequency (DIF) FFT algorithm. Finally in Section 8.5, the inverse DFT is introduced and its implementation is discussed.

8.2 THE DISCRETE FOURIER TRANSFORM

The frequency analysis of a sampled signal $x(n)$ is performed using the Discrete-Time Fourier Transform (DTFT) $X(e^{j\omega})$ given by

$$X(e^{j\omega}) = \sum_n x(n)e^{-j\omega n}$$

However, the DTFT is a continuous function of ω. Therefore its numerical evaluation requires sampling in the frequency domain. If the sequence $x(n)$ is of finite length N, then N samples of its DTFT $X(e^{j\omega})$ over 0 to 2π interval completely describe the sequence $x(n)$. This sampled version of $X(e^{j\omega})$ is evaluated using the discrete Fourier Transform (DFT) $X(k)$ given by

$$X(k) = \sum_{n=0}^{N-1} x(n)e^{-j\frac{2\pi}{N}nk} \quad , \quad k = 0, \ldots, N-1$$

To simplify the notation, it is desirable to define the complex-valued phase factor W_N, which is the Nth root of unity, as

$$W_N = e^{-j\,2\pi/N}$$

Then the DFT equation becomes

$$X(k) = \sum_{n=0}^{N-1} x(n)W_N^{nk} \quad , \quad k = 0, \ldots, N-1 \tag{8-1}$$

A complex summation of N complex multiplications ($4N$ real multiplications) and N-1 complex additions ($4N$-2 real additions) is required for each of the N output samples. Consequently, to compute all N values of the DFT requires N^2 complex multiplications and $N^2 - N$ complex additions. This direct computation of the DFT is basically inefficient primarily because it does not exploit the symmetry and periodicity properties of the phase factor W_N. The time burden created by this large number of calculations limits the usefulness of the DFT in many applications. For this reason, tremendous amount of a effort was devoted to developing more efficient ways of computing the DFT. This effort produced computationally efficient algorithms which are collectively known as fast Fourier transform (FFT) algorithms.

There are two different approaches to efficiently computing the DFT. In the first approach, a divide-and-conquer strategy is employed in which an N-point DFT, where N is a composite number, is reduced to the computation of smaller DFTs from which the larger DFT is computed. In this chapter, we will present this approach for computing the DFT when the size N is a power of 2. In particular, we will describe two important algorithms, the radix-2 decimation-in-time fast Fourier transform (DIT FFT) and the radix-2 decimation-in-frequency fast Fourier transform (DIF FFT), and their implementation on the ADSP-2101.

The second approach is based on the formulation of the DFT as a linear filtering operation on the data. This approach leads to two algorithms, the Goertzel algorithm and the chirp-z transform algorithm. The Goertzel algorithm is described in Section 9.6.

8.3 RADIX-2 DECIMATION-IN-TIME FFT

The decimation-in-time (DIT) FFT divides the input (time) sequence into two groups, one of even samples and the other of odd samples. N/2-point DFTs are performed on these sub-sequences, and their outputs are combined to form the N-point DFT.

8.3.1 The Algorithm

The decimation-in-time methodology is illustrated by the following equations (Proakis and Manolakis [8]). First, $x(n)$, the input sequence in equation (8-1), is divided into even and odd sub-sequences:

$$X(k) \quad = \quad \sum_{n=0}^{\frac{N}{2}-1} x(2n)W_N^{2nk} + \sum_{n=0}^{\frac{N}{2}-1} x(2n+1)W_N^{(2n+1)k}$$

$$= \quad \sum_{n=0}^{\frac{N}{2}-1} x(2n)W_N^{2nk} + W_N^k \sum_{n=0}^{\frac{N}{2}-1} x(2n+1)W_N^{2nk}, k = 0,...,N-1 \qquad (8-2)$$

By the substitutions

$$W_N^{2nk} = (e^{-j2\pi/N})^{2nk} = (e^{-j2\pi/(N/2)})^{nk} = W_{N/2}^{nk}$$

$$x_1(n) = x(2n)$$

$$x_2(n) = x(2n+1)$$

this equation becomes

$$X(k) \quad = \quad \sum_{n=0}^{\frac{N}{2}-1} x_1(n)W_{N/2}^{nk} + W_N^k \sum_{n=0}^{\frac{N}{2}-1} x_2(n)W_{N/2}^{nk}$$

$$= \quad Y(k) + W_N^k Z(k) \quad , \quad k = 0,...,N-1 \qquad (8-3)$$

Equation (8-3) is the sum of two $N/2$-point DFTs ($Y(k)$ and $Z(k)$) performed on the sub-sequences of even and odd samples, respectively, of the input sequence, $x(n)$. Multiples of W_N (called "twiddle factors") appear as coefficients in the FFT calculation. In equation (8-3), $Z(k)$ is multiplied by the twiddle factor W_N^k.

Because $W_N^{k+N/2} = (e^{-j2\pi/N})^k \cdot (e^{-j2\pi/N})^{N/2} = -W_N^k$, equation (8-3) can also be expressed as two equations:

$$X(k) \quad = \quad Y(k) + W_N^k Z(k)$$

$$X(k+N/2) \quad = \quad Y(k) - W_N^k Z(k) \quad , \quad k = 0,...,\frac{N}{2}-1 \qquad (8-4)$$

Together these equations form an N-point FFT. Figure 8-1 illustrates this first decimation of the DFT.

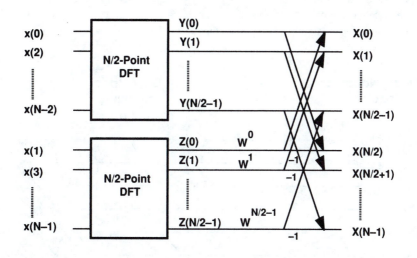

Figure 8-1:   First Decimation of DIT FFT

The two $N/2$-point DFTs ($Y(k)$ and $Z(k)$) can be divided to form four $N/4$-point DFTs, yielding equation pairs (8-5) and (8-6).

$$Y(k) \quad = \quad U(k) + W_N^{2k} V(k)$$

$$Y(k+N/4) \quad = \quad U(k) - W_N^{2k} V(k) \quad , \quad n = 0, \ldots, \frac{N}{4} - 1 \tag{8-5}$$

$$Z(k) \quad = \quad R(k) + W_N^{2k} S(k)$$

$$Z(k+N/4) \quad = \quad R(k) - W_N^{2k} S(k) \quad , \quad n = 0, \ldots, \frac{N}{4} - 1 \tag{8-6}$$

$U(k)$ and $V(k)$ are $N/4$-point DFTs whose input sequences are created by dividing $x_1(n)$ into even and odd sub-sequences. Similarly, $R(k)$ and $S(k)$ are $N/4$-point DFTs performed on the even and odd sub-sequences of $x_2(n)$. Each of these four equations can be divided to form two more. The final decimation occurs when each pair of equations together computes a two-point DFT (one point per equation). The pair of equations that make up the two-point DFT is called a radix-2 "butterfly." The butterfly is the core calculation of the FFT. The entire FFT is performed by combining butterflies in patterns determined by the FFT algorithm.

A complete eight-point DIT FFT is illustrated graphically in Figure 8-2. Each pair of arrows represents a butterfly. Notice that the entire FFT computation is made up of butterflies organized in different patterns, called groups and stages. The first stage consists of four groups of one butterfly each. The second stage has two groups of two butterflies, and the last has one group of four butterflies. Every stage contains $N/2$ (four) butterflies. Each butterfly has two input points, called the dual node and the primary node. The spacing between the nodes in the sequence is called the dual-node spacing. Associated with each butterfly is a twiddle factor whose exponent depends on the group and stage of the butterfly.

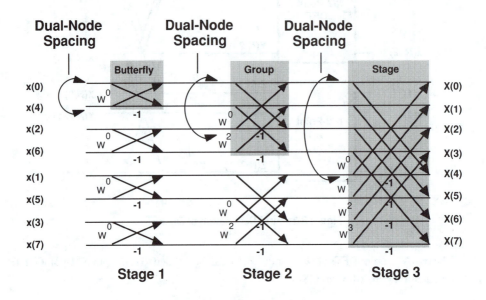

Figure 8-2: Eight-Point DIT FFT

Notice that whereas the output sequence is sequentially ordered, the input sequence is not. This is an effect of repeatedly dividing the input sequence into sub-sequences of even and odd samples. It is possible to perform an FFT using input and output sequences in other orders, but these approaches generally complicate addressing in the FFT program and can require a different butterfly. In this section, we have opted to scramble the input sequence of the DIT FFT because this approach uses twiddle factors in sequential order, produces the output sequence in sequential order, and requires a relatively simple butterfly. The scrambling of the inputs is achieved by a process called bit reversal, which is described later in this chapter.

A generalized butterfly flow graph is shown in Figure 8-3. The variables x and y represent the real and imaginary parts, respectively, of a sample. The twiddle factor can be divided into real and imaginary parts because $W_N = e^{-j2\pi/N} = \cos(2\pi/N) - j\sin(2\pi/N)$. In the program presented later in this section, the twiddle factors are initialized in memory as cosine and -sine values (not +sine). For this reason, the twiddle factors are shown in Figure 8-3 as $C + j(-S)$. C represents cosine and $-S$ represents -sine.

The dual node $(x_1 + jy_1)$ is multiplied by the twiddle factor $C + j(-S)$. The result of this multiplication is added to the primary node $(x_0 + jy_0)$ to produce $x_0' + jy_0'$ and subtracted from the primary node to produce $x_1' + jy_1'$. Equations (8-7) through (8-10) calculate the real and imaginary parts of the butterfly outputs.

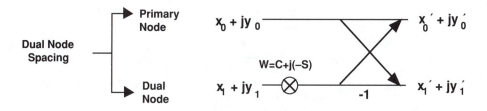

Figure 8-3: Radix-2 DIT FFT Butterfly

$$x_0' = x_0 + (C)x_1 - (-S)y_1 \qquad\qquad (8-7)$$

$$y_0' = y_0 + (C)y_1 + (-S)x_1 \qquad\qquad (8-8)$$

$$x_1' = x_0 - (C)x_1 - (-S)y_1 \qquad\qquad (8-9)$$

$$y_1' = y_0 - (C)y_1 + (-S)x_1 \qquad\qquad (8-10)$$

The butterfly produces two complex outputs that become butterfly inputs in the next stage of the FFT. Because each stage has the same number of butterflies ($N/2$), the number of butterfly inputs and outputs remains the same from one stage to the next. An "in-place" implementation writes each butterfly output over the corresponding butterfly input (x_0' overwrites x_0, etc.) for each butterfly in a stage. In an in-place implementation, the FFT results end up in the same memory range as the original inputs.

8.3.2 A DIT FFT Program

The flow chart for the DIT FFT program is shown in Figure 8-4. The FFT program is divided into three subroutines. The first subroutine scrambles the input data. The next subroutine computes the FFT, and the third scales the output data.

Four modules are created. The main module declares and initializes data buffers and calls subroutines. The other three modules contain the FFT, bit reversal, and block floating-point scaling subroutines. The main module and the FFT module are described in this section. The bit reversal and block floating-point scaling modules are described later in this section.

Main Module

The *dit_fft_main* module is shown in Listing 8-1. N is the number of points in the FFT (in this example, $N = 1024$) and N_div_2 is used for specifying the lengths of buffers. The number of points in the FFT can be changed by changing the value of these constants and the twiddle factors. The data buffers *twid_real* and *twid_imag* in program memory hold the twiddle factor cosine and sine values. The *inplacereal, inplaceimag, inputreal* and *inputimag* buffers in data memory store real and imaginary data values. Sequentially ordered input data is stored in *inputreal* and *inputimag*. This data is scrambled and written to *inplacereal* and *inplaceimag*, which are the data buffers used by the in-place FFT. A four-location buffer called *padding* is placed at the end of *inplaceimag* to allow data accesses to exceed the buffer length. If no padding

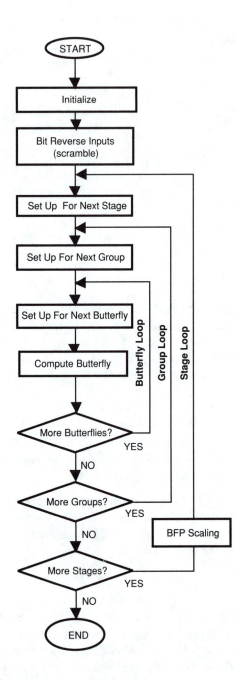

Figure 8-4: Radix-2 DIT FFT Flow Chart

was placed after *inplaceimag* and the program attempted to read undefined memory locations, the ADSP-2101 Simulator would signal an error. This buffer assists in debugging but is not necessary in a real system. Variables (one-location buffers) *groups*, *bflys_per_group*, *blk_exponent* and *node_space* are declared last.

The real part (cosine values) of the twiddle factors in the *twid_real.dat* file are placed in the buffer *twid_real*. Likewise, *twid_imag.dat* is placed in *twid_imag*. The variable *groups* is initialized to *N_div_2*, and *bflys_per_group* and *node_space* are initialized to one because there are *N*/2 groups of one butterfly in the first stage of the FFT. The *blk_exponent* is initialized to zero. This exponent value is updated when the output data is scaled.

Two subroutines are called. The first subroutine places the input sequence in bit-reversed order. The second performs the FFT and calls the block floating-point scaling routine.

```
{ DIT FFT Main program                                        DIT_MAIN.DSP

    This program computes a 1024-point DFT of data values stored data
    buffers using DIT FFT algorithm.  The input data is assumed to be
    complex.  The real and imaginary input data are stored in inputreal
    and inputimag buffers, respecively.  The input data is read from the
    disk files INPUTREA.DAT and INPUTIMA.DAT. The real and imaginary parts
    of twiddle factors are read from disk files TWID_REA.DAT and TWID_
    IMA.DAT, respectively.  The output DFT values are available in data
    memory locations inplacereal and inplaceimg.
}

.MODULE/ABS=4            dit_fft_main;
.CONST                   N=1024, N_div_2=512;   {Const. for 1024 points}
.VAR/PM/RAM/CIRC         twid_real [N_div_2];
.VAR/PM/RAM/CIRC         twid_imag [N_div_2];
.VAR/DM/RAM/ABS=0        inplacereal [N], inplaceimag [N];
.VAR/DM/RAM/ABS=0x1000   inputreal [N], inputimag [N], padding [4];
.VAR/DM/RAM              groups, bflys_per_group,
                         node_space, blk_exponent;

.INIT     twid_real: <twid_real.dat>;
.INIT     twid_imag: <twid_imag.dat>;
.INIT     inputreal: <inputreal.dat>;
.INIT     inputimag: <inputimag.dat>;
.INIT     inplaceimag: <inputimag.dat>;
.INIT     groups: N_div_2;
.INIT     bflys_per_group: 1;
.INIT     node_space: 1;
.INIT     blk_exponent: 0;
.INIT     padding: 0,0,0,0;                     {Zeros after inplaceimag}

.GLOBAL   inplacereal, inplaceimag;
.GLOBAL   inputreal, inputimag;
.GLOBAL   twid_real, twid_imag;
.GLOBAL   groups, bflys_per_group, node_space, blk_exponent;

.EXTERNAL scramble, fft_strt;
          CALL scramble;                        {Subroutine calls}
          CALL fft_strt;
          TRAP;                                 {Halt program}
.ENDMOD;
```

———— Listing 8-1: Main Module, Radix-2 DIT FFT ————

DIT FFT Module

The FFT routine uses three nested loops. The inner loop computes butterflies, the middle loop controls the grouping of these butterflies, and the outer loop controls the FFT stage characteristics. These loops are described separately in the following sections. The complete routine is presented in Listing 8-7.

Butterfly Loop

The butterfly calculation involves a complex multiplication, a complex addition, and a complex subtraction. These operations can potentially cause the butterfly data to grow by two bits from input to output. For example, if x_0 is 0x07FF (five sign bits), x_0' could be 0x100F (three sign bits). Because of this bit growth, precautions must be taken to ensure that 16-bit data never overflows.

An example of bit growth and overflow is shown below.

Bit Growth:
Input to butterfly 0x0F00   = 0000 1111 0000 0000
Output from butterfly 0x1E00   = 0001 1110 0000 0000

Overflow:
Input to butterfly 0x7000   = 0111 0000 0000 0000
Output from butterfly 0xE000   = 1110 0000 0000 0000

In overflow, the positive number 0x7000 is multiplied by a positive number, resulting in

0xE000, which is too large to represent as a positive, signed 16-bit number. 0xE000 is erro

neously interpreted as a negative number.

To avoid errors caused by overflow, one of three methods of compensating for bit growth can be applied:

- Input data scaling
- Unconditional block floating-point scaling (output data)
- Conditional block floating-point scaling (output data)

Three different code segments for the butterfly calculation are presented in this section; each uses a different method of compensating for bit growth.

One way to ensure that overflow never occurs is to include enough extra sign bits, called guard bits, in the FFT input data to ensure that bit growth never results in overflow (Rabiner and Gold, [11]). Data can grow by a maximum factor of 2.4 from butterfly input to output (two bits of growth). However, a data value cannot grow by this maximum amount in two consecutive stages. The number of guard bits necessary to compensate for the maximum possible bit growth in an N-point FFT is $\log_2 N + 1$. For example, each of the input samples of a 32-point FFT (requiring five stages) must contain six guard bits, so ten bits are available for data (one sign bit, nine magnitude bits). This method requires no data shifting and is therefore the fastest of the three methods discussed in this section. However, for large FFTs the resolution of the input data is greatly limited. For small, low-precision FFTs, this is the fastest and most efficient method.

The code segment for a butterfly with no shifting is shown in Listing 8-2. This section of code computes one butterfly equation while setting up values for the next butterfly. The butterfly outputs (x_0', y_0', x_1' and y_1') are written over the inputs to the butterfly (x_0, y_0, x_1 and y_1') in the boldface instructions. The input and output parameters of the butterfly loop are shown below.

Initial Conditions

$MX0 = x_1$
$MY0 = C$
$MY1 = (-S)$
$I0 \rightarrow x_0$
$I1 \rightarrow x_1$
$I2 \rightarrow y_0$
$I3 \rightarrow y_1$
$I4 \rightarrow$ next C
$I5 \rightarrow$ next $(-S)$
$I6 \rightarrow y_1$
$CNTR =$ butterfly count
$M0 = 0$
$M1 = 1$
$M4 =$ twiddle factor modify value
$M5 = 1$

Final Conditions

$MX0 =$ next x_1
$MY0 =$ next C
$MY1 =$ next $(-S)$
$I0 \rightarrow$ next x_0
$I1 \rightarrow$ next x_1
$I2 \rightarrow$ next y_0
$I3 \rightarrow$ next y_1
$I4 \rightarrow C$ after next C
$I5 \rightarrow (-S)$ after next $(-S)$
$I6 \rightarrow$ next y_1
$CNTR =$ butterfly count -1

```
MR=MX0*MY0(SS),MX1=DM(I6,M5);        {MR=x1(C),MX1=y1}
MR=MR-MX1*MY1(RND),AY0=DM(I0,M0);    {MR=x1(C)-y1(-S),AY0=x0}
AR=MR1+AY0,AY1=DM(I2,M0);            {AR=x0+[x1(C)-y1(-S)],AY1=y0}
AR=AY0-MR1,DM(I0,M1)=AR;             {AR=x0-[x1(C)-y1(-S)]},
                                     { x0´=x0+[x1(C)-y1(-S)]}

MR=MX0*MY1(SS),DM(I1,M1)=AR;         {MR=x1(-S),x1´=x0-[x1(C)-y1(-S)]}
MR=MR+MX1*MY0(RND),MX0=DM(I1,M0),MY1=PM(I5,M4);
                                     {MR=x1(-S)+y1(C),MX0=next x1,}
                                     { MY1=next (-S)}
AR=AY1-MR1,MY0=PM(I4,M4);            {AR=y0-[x1(-S)+y1(C)],MY0=next C}
AR=MR1+AY1,DM(I3,M1)=AR;             {AR=y0+[x1(-S)+y1(C)],}
                                     { y1´=y0-[x1(-S)+y1(C)]}

DM(I2,M1)=AR;                        {y0´=y0+[x1(-S)+y1(C)]}
```

——————————— **Listing 8-2: DIT FFT Butterfly, Input Data Scaled** ———————————

Another way to compensate for bit growth is to scale the outputs down by a factor of two unconditionally after each stage. This approach is called unconditional block floating-point scaling. Initially, two guard bits are included in the input data to accommodate the maximum bit growth in the first stage. In each butterfly of a stage calculation, the data can grow into the guard bits. To prevent overflow in the next stage, the guard bits are replaced before the next stage is executed by shifting the entire block of data one bit to the right and updating the block exponent. This shift is necessary after every stage except the last, because no overflow can occur after the last stage.

The input data to an unconditional block floating-point FFT can have at most 14 bits (one sign bit and 13 magnitude bits). In the FFT calculation, the data loses a total of ($\log_2 N$)-1 bits because of shifting. Unconditional block floating-point scaling results in the same number of bits lost as in input data scaling. However, it produces more precise results because the FFT starts with more precise input data. The tradeoff is a slower FFT calculation because of the extra cycles needed to shift the output of each stage.

The code for the unconditional block floating-point butterfly is shown in Listing 8-3. Instructions that write butterfly results to memory are boldface. After the last stage of the FFT, no compensation for bit growth is needed, so a butterfly with no shifting can be used in the last stage.

Initial Conditions	*Final Conditions*
SR0 = last y_0'	SR0 = y_0'
MX0 = x_1	MX0 = next x_1
MX1 = y_1	MX1 = next y_1
MY0 = C	MY0 = next C
MY1 = $(-S)$	MY1 = next $(-S)$
I0 → x_0	I0 → next x_0
I1 → x_1	I1 → next x_1
I2 → last y_0'	I2 → y_0'
I3 → y_1	I3 → next y_1
I4 → next C	I4 → C after next C
I5 → next $(-S)$	I5 → $(-S)$ after next $(-S)$
I6 → next y_1	I6 → y_1 after next y_1
CNTR = butterfly count	CNTR = butterfly count -1
M0 = 0	
M1 = 1	
M4 = twiddle factor modify value	
M5 = 1	
SE = -1	

```
MR=MX0*MY0(SS),DM(I2,M1)=SR0;                    {MR=x1(C),last y0=last y0´}
MR=MR-MX1*MY1(RND),AY0=DM(I0,M0);               {MR=x1(C)-y1(-S),AY0=x0}
AR=MR1+AY0,AY1=DM(I2,M0);                        {AR=x0+[x1(C)-y1(-S)],AY1=y0}
SR=ASHIFT AR(LO);                               {Shift result right 1 bit}
DM(I0,M1)=SR0,AR=AY0-MR1;                        {x0´=x0-[x1(C)-y1(-S)],}
                                                { AR=x0-[x1(C)-y1(-S)]}

SR=ASHIFT AR(LO);                               {Shift result right 1 bit}
DM(I1,M1)=SR0,MR=MX0*MY1(SS);                    {x1´=x0-[x1(C)-y1(-S)],MR=x1(-S)]}
MR=MR+MX1*MY0(RND),MX0=DM(I1,M0),MY1=PM(I5,M4);
                                                {MR=x1(-S)-y1(C),MX0=next}
                                                { x1,MY1=next(-S)}
AR=AY1-MR1,MY0=PM(I4,M4);                        {AR=y0-[x1(-S)-y1(C)],MY0=next C}
SR=ASHIFT AR(LO),MX1=DM(I6,M5);                  {Shift result right 1 bit,}
                                                { MX1=next y1}
DM(I3,M1)=SR0,AR=MR1+AY1;                        {y1´=y0-[x1(-S)-y1(C),}
                                                { AR=y0+[x1(-S)-y1(C)]}
SR=ASHIFT AR(LO);                               {Shift result right 1 bit}
```

─────────── **Listing 8-3: DIT FFT Butterfly, Unconditional Block Floating-Point Scaling** ───────────

In conditional block floating-point scaling, data is shifted only if bit growth occurs. If one or more outputs grows, the entire block of data is shifted to the right and the block exponent is updated. For example, if the original block exponent is 0 and data is shifted three positions, the resulting block exponent is +3.

The code segment for the conditional block floating-point butterfly is shown in Listing 8-4. As in the other types of butterflies, one butterfly equation is calculated and its outputs (x_0', y_0', x_1' and y_1') are written over its inputs (x_0, y_0, x_1 and y_1) in the boldface instructions.

The conditional block floating-point butterfly checks each butterfly output for growth with the EXPADJ instruction. This instruction does no shifting; instead, it monitors the output data and updates the SB register if bit growth is detected. (See the ADSP-2101 User's Manual for a complete description of this instruction.) If shifting is necessary it is performed after the entire stage is complete (in the block floating-point scaling routine). The butterfly code computes one butterfly equation while setting up values for the next butterfly. The input and output parameters of the butterfly loop are as follows:

Initial Conditions	*Final Conditions*
MX0 = x_1	MX0 = next x_1
MX1 = y_1	MX1 = next y_1
MY0 = C	MY0 = next C
MY1 = ($-S$)	MY1 = next ($-S$)
I0 x_0	I0 next x_0
I1 -> x_1	I1 next x_1
I2 y_0	I2 next y_0
I3 y_1	I3 next y_1
I4 next C	I4 C after next C
I5 next ($-S$)	I5 ($-S$) after next ($-S$)
CNTR = butterfly count	CNTR = butterfly count -1
M1 = 1	
M4 = twiddle factor modify value	
M0 = 0	
SB = monitored block exponent for this stage	

```
MR=MX0*MY1(SS),AX0=DM(I0,M0);            {MR=x1(-S),AX0=x0}
MR=MR+MX1*MY0(RND),AX1=DM(I2,M0);        {MR=[y1(C)+x1(-S)];AX1=y0}
AY1=MR1,MR=MX0*MY0(SS);                  {AY1=[y1(C)+x1(-S)];MR=x1(C)}
MR=MR-MX1*MY1(RND);                      {MR=[x1(C)-y1(-S)]}
AY0=MR1,AR=AX1-AY1;                      {AY0=[x1(C)-y1(-S)],}
                                         { AR=y0-[y1(C)+x1(-S)]}

SB=EXPADJ AR,DM(I3,M1)=AR;               {check for bit growth,}
                                         { y1´=y0-[y1(C)+x1(-S)]}

AR=AX0-AY0,MX1=DM(I3,M0),MY1=PM(I5,M4);
                                         {AR=x0-[x1(C)-y1(-S)],MX1=next y1,}
                                         { MY1=next S}

SB=EXPADJ AR,DM(I1,M1)=AR;               {check for bit growth,}
                                         { x1´=x0-[x1(C)-y1(-S)]}

AR=AX0+AY0,MX0=DM(I1,M0),MY0=PM(I4,M4);
                                         {AR=x0+[x1(C)-y1(-S)],MX0=next x1,}
                                         { MY0=next C}

SB=EXPADJ AR,DM(I0,M1)=AR;               {check for bit growth,}
                                         { x0´=x0+[x1(C)-y1(-S)]}

AR=AX1+AY1;                              {AR=y0+[y1(C)+x1(-S)]}
SB=EXPADJ AR,DM(I2,M1)=AR;               {check for bit growth,}
                                         { y0´=y0+[y1(C)+x1(-S)]}
```

——————— **Listing 8-4: DIT FFT Butterfly, Conditional Block Floating-Point Scaling** ———————

Group Loop

The group loop controls the grouping of butterflies. It sets pointers to the input data and twiddle factors of the first butterfly in the group, initializes the butterfly counter and sets up the butterfly loop for each group.

The code segment for the group loop is shown in Listing 8-5. This code is designed for the conditional block floating-point butterfly and thus requires slight modification for use with the other types (input scaling, unconditional block floating-point) of butterflies. The first butterfly of every group in the first stage of the DIT FFT has a twiddle factor of W^0. Thus, I4 and I5 are initialized to point to the cosine and sine values of W^0 before the butterfly loop is entered. In the group loop, the butterfly counter is initialized and initial butterfly data is fetched. The butterfly loop is executed bflys_per_group times to compute all butterflies in the group. After the butterfly loop is complete, pointers I0, I1, I2 and I3 are updated with the MODIFY instruction to point to x_0, x_1, y_0 and y_1 of the first butterfly in the next group. The group loop is executed groups times.

The input and output parameters of the group loop are as follows:

Initial Conditions	*Final Conditions*
I0 $\rightarrow x_0$ of first butterfly in group	I0 $\rightarrow x_0$ of first butterfly in next group
I1 $\rightarrow x_1$ of first butterfly in group	I1 $\rightarrow x_1$ of first butterfly in next group
I2 $\rightarrow y_0$ of first butterfly in group	I2 $\rightarrow y_0$ of first butterfly in next group
I3 $\rightarrow y_1$ of first butterfly in group	I3 $\rightarrow y_1$ of first butterfly in next group
CNTR = group count	CNTR = group count -1
M2 = node_space	

```
            I4=^twid_real;
            I5=^twid_imag;              {Initialize twiddle factor pointers}
            CNTR=DM(bflys_per_group);   {Initialize butterfly counter}
            MY0=PM(I4,M4),MX0=DM(I1,M0);      {MY0=C,MX0=x1}
            MY1=PM(I5,M4),MX1=DM(I3,M0);      {MY1=(-S),MX1=y1}
            DO bfly_loop UNTIL CE;

bfly_loop:      {Calculate All Butterflies in Group}

            MODIFY(I0,M2);      {I0 first x0 in next group}
            MODIFY(I1,M2);      {I1 first x1 in next group}
            MODIFY(I2,M2);      {I2 first y0 in next group}
group_loop: MODIFY(I3,M2);      {I3 first y1 in next group}
```

———————————— **Listing 8-5: Radix-2 DIT FFT Group Loop** ————————————

Stage Loop

The stage loop controls the grouping characteristics of the FFT. These include the number of groups in a stage, the number of butterflies in each group, and the node spacing. The stage loop also calls a subroutine which performs conditional block floating-point scaling on the outputs of a stage calculation. Note that if unconditional block floating-point scaling or input data scaling were used, this call would be omitted.

The stage loop code for a conditional block floating-point FFT is shown in Listing 8-6. The stage loop sets up the group loop by initializing I0, I1, I2 and I3 to point to x_0, x_1, y_0 and y_1, respectively, for the first butterfly in the first group. It also initializes the group loop counter and node space modifier so that pointers can be updated for new groups. The value of the twiddle factor exponent is increased by groups for each butterfly. M4, initialized to groups, is the modifier for the twiddle factor pointers.

The group loop calculates all groups in the stage. After the group loop is complete, a block floating-point subroutine is called to check the stage outputs for bit growth and scale the data if necessary. The stage characteristics are then updated for the next stage; *bflys_per_group* and *node_space* are doubled and groups is divided by two.

The input and output parameters for the stage loop are as follows. Note that all the parameters except the stage count are passed in memory.

Initial Conditions	*Final Conditions*
groups=# groups current stage	*groups*=# groups next stage
bflys_per_group=# butterflies/group	*bflys_per_group*=# butterflies/ group next stage
node_space=node spacing current stage	*node_space*=node spacing next stage
CNTR=stage count	CNTR=stage count -1
inplacereal=real stage input data	*inplacereal*=real stage output data
inplaceimag=imag. stage input data	*inplaceimag*=imag. stage output data

```
        I0=^inplacereal;            {I0 first x0 in first group of stage}
        I2=^inplaceimag;            {I2 first y0 in first group of stage}
        SB=-2                       {SB = -(number of guard bits)}
        SI=DM(groups);              {SI = groups}
        CNTR=SI;                    {Initialize group counter}
        M4=SI;                      {Initialize twiddle factor modifier}
        M2=DM(node_space);          {Initialize node spacing modifier}
        I1=I0;
        MODIFY(I1,M2);              {I1 first x1 of first group in stage}
        I3=I2;
        MODIFY(I3,M2);              {I3 first y1 of first group in stage}
        DO group_loop UNTIL CE;

group_loop:                         {Compute All Groups in Stage}

        CALL bfp_adj;               {Adjust stage output for bit growth}
        SI=DM(bflys_per_group);
        SR=ASHIFT SI BY 1(LO);
        DM(node_space)=SR0;         {node_space=node_space ¥ 2}
        DM(bflys_per_group)=SR0;    {bflys_per_group=bflys_per_group ¥ 2}
        SI=DM(groups);
        SR=ASHIFT SI BY -1(LO);
        DM(groups)=SR0;             (groups = groups ÷ 2)
```

—————————— **Listing 8-6: Radix-2 DIT FFT Stage Loop** ——————————

DIT FFT Subroutine

The complete conditional block floating-point radix-2 DIT FFT routine is shown in Listing 8-7. The constants N and $\log_2 N$ are the number of points and the number of stages in the FFT, respectively. To change the number of points in the FFT, you modify these constants. Notice that the length and modify registers (that retain the same values throughout the FFT calculation) and the stage counter are initialized before the stage loop is executed. Instructions that write butterfly results to memory are boldface.

```
{ Radix-2 DIT FFT Subroutine                                    R2DITFFT.DSP

     Performs Radix-2 DIT FFT

     Calling Parameters
         inplacereal = Real input data in scrambled order
         inplaceimag = All zeroes (real input assumed)
         twid_real = Twiddle factor cosine values
         twid_imag = Twiddle factor sine values
         groups = N/2
         bflys_per_group = 1
         node_space = 1

     Return Values
         inplacereal = Real FFT results in sequential order
         inplaceimag = Imaginary FFT results in sequential order

     Altered Registers
         I0,I1,I2,I3,I4,I5,L0,L1,L2,L3,L4,L5
         M0,M1,M2,M3,M4,M5
         AX0,AX1,AY0,AY1,AR,AF
         MX0,MX1,MY0,MY1,MR,SB,SE,SR,SI

     Altered Memory
         inplacereal, inplaceimag, groups, node_space,
         bflys_per_group, blk_exponent
}
.MODULE    radix2_dit_fft;
.CONST     logN=10, N=1024;          {Set constants for N-point FFT}
.EXTERNAL  twid_real, twid_imag;
.EXTERNAL  inplacereal, inplaceimag;
.EXTERNAL  groups, bflys_per_group, node_space;
.EXTERNAL  bfp_adj;
.ENTRY     fft_strt;

fft_strt:  CNTR=logN;                {Initialize stage counter}
           M0=0;
           M1=1;
           L1=0;
           L2=0;
           L3=0;
           L4=%twid_real;
           L5=%twid_imag;
           DO stage_loop UNTIL CE;    {Compute all stages in FFT}
               I0=^inplacereal;       {I0 x0 in 1st grp of stage}
               I2=^inplaceimag;       {I2 y0 in 1st grp of stage}
               SB=-2                   {SB to detect data > 14 bits}
               SI=DM(groups);
               CNTR=SI;                {CNTR = group counter}
               M4=SI;                  {M4=twiddle factor modifier}
               M2=DM(node_space);      {M2=node space modifier}
               I1=I0;
               MODIFY(I1,M2);          {I1 x1 of 1st grp in stage}
               I3=I2;
               MODIFY(I3,M2);          {I3 y1 of 1st grp in stage}
```

```
            DO group_loop UNTIL CE;
                I4=^twid_real;        {I4   C of W⁰}
                I5=^twid_imag;        {I5   (-S) of W⁰}
                CNTR=DM(bflys_per_group); {CNTR = butterfly counter}
                MY0=PM(I4,M4),MX0=DM(I1,M0); {MY0=C,MX0=x1 }
                MY1=PM(I5,M4),MX1=DM(I3,M0); {MY1=-S,MX1=y1}
                DO bfly_loop UNTIL CE;
                    MR=MX0*MY1(SS),AX0=DM(I0,M0); {MR=x1(-S),AX0=x0}
                    MR=MR+MX1*MY0(RND),AX1=DM(I2,M0);
                                      {MR=(y1(C)+x1(-S)),AX1=y0}
                    AY1=MR1,MR=MX0*MY0(SS); {AY1=y1(C)+x1(-S),MR=x1(C) }
                    MR=MR-MX1*MY1(RND);   {MR=x1(C)-y1(-S) }
                    AY0=MR1,AR=AX1-AY1;   {AY0=x1(C)-y1(-S),}
                                          {AR=y0-[y1(C)+x1(-S)]}
                    SB=EXPADJ AR,DM(I3,M1)=AR; {Check for bit growth,}
                                              {y1´=y0-[y1(C)+x1(-S)]}
                    AR=AX0-AY0,MX1=DM(I3,M0),MY1=PM(I5,M4);
                                          {AR=x0-[x1(C)-y1(-S)],}
                                          {MX1=next y1,MY1=next (-S)}
                    SB=EXPADJ AR,DM(I1,M1)=AR; {Check for bit growth,}
                                              {x1´=x0-[x1(C)-y1(-S)]}
                    AR=AX0+AY0,MX0=DM(I1,M0),MY0=PM(I4,M4);
                                          {AR=x0+[x1(C)-y1(-S)],}
                                          {MX0=next x1,MY0=next C}
                    SB=EXPADJ AR,DM(I0,M1)=AR; {Check for bit growth,}
                                              {x0´=x0+[x1(C)-y1(-S)]}
                    AR=AX1+AY1;           {AR=y0+[y1(C)+x1(-S)]}
bfly_loop:          SB=EXPADJ AR,DM(I2,M1)=AR; {Check for bit growth,}
                                              {y0´=y0+[y1(C)+x1(-S)]}
                MODIFY(I0,M2);        {I0 1st x0 in next group}
                MODIFY(I1,M2);        {I1 1st x1 in next group}
                MODIFY(I2,M2);        {I2 1st y0 in next group}
group_loop:     MODIFY(I3,M2);        {I3 1st y1 in next group}
            CALL bfp_adj;             {Compensate for bit growth}
            SI=DM(bflys_per_group);
            SR=ASHIFT SI BY 1(LO);
            DM(node_space)=SR0;       {node_space=node_space × 2}
            DM(bflys_per_group)=SR0;
{bflys_per_group= }
                                      {bflys_per_group × 2}
            SI=DM(groups);
            SR=ASHIFT SI BY -1(LO);
stage_loop: DM(groups)=SR0;           {groups=groups ÷ 2}
        RTS;
.ENDMOD;
```

—————————— **Listing 8-7: Radix-2 DIT FFT Routine, Conditional Block Floating-Point** ——————————

Bit Reversal

Bit reversal is an addressing technique used in FFT calculations to obtain results in sequential order. Because the FFT repeatedly subdivides data sequences, the data and/or twiddle factors may be scrambled (in bit-reversed order). All radix-2 FFTs can be calculated with either the input sequence or the output sequence scrambled. The twiddle factors may also need to be scrambled, depending on the order of the input and output sequences. In this chapter, however, input and output sequences are set up so that twiddle factors are never scrambled. This simplifies the FFT explanation as well as the program.

As described earlier, the input sequence to the radix-2 DIT FFT is scrambled before the FFT is performed. This scrambling is accomplished through bit reversal. Bit reversal operates on the binary number that represents the position of a sample within an array of samples. The bit-reversed position is the transpose of the bits of the binary number about its center; for example the transpose of the 3-bit binary number 100 is 001. (In this example, three bits represent eight positions, so bits zero and two are interchanged.) Four bits are needed to represent 16 positions, so in a 16-point sequence, bits zero and three and bits one and two would be interchanged. A 1024-point sequence requires the reversal of ten bits.

The ADSP-2101 has a bit-reverse capability built into its data address generator #1 (DAG1). When a mode bit is enabled (through software), the 14-bit address generated by DAG1 is automatically bit-reversed for any data memory read or write. The two address generators of the ADSP-2101 greatly simplify bit reversal. One address generator can be used to read sequentially ordered data, and the other can be used to write the same data to its bit-reversed location. Because the address generators are independent, intermediate enabling and disabling of the bit-reverse mode is not needed.

In many cases, fewer than 14 bits must be reversed (for example, an eight-point FFT needs only three bits reversed). Reversal of fewer than 14 bits is accomplished by adding the correct modify value to the address pointer after each memory access. The following example demonstrates bit reversal of ten bits using I0 to store the address to be reversed and M0 to store the modify value.

First, we determine the first bit-reversed address. This address is the first 14-bit address with the ten least significant bits reversed. For the DIT FFT subroutine, the first address in the *inplacereal* buffer is 0x0000. If we reverse the ten least significant bits of 0x0000, we still have 0x0000. Thus, we want to output 0x0000 as the first bit-reversed address. To do so, I0 must be initialized to the number that, when bit-reversed by the ADSP-2101 (all 14 bits), is 0x0000. In this case, that number is also 0x0000.

The second bit-reversed address must be 0x0200 (0x0001 with ten least significant bits reversed). We must modify I0 to the value that, when bit-reversed (all 14 bits) is 0x0200. This value is 0x0010. Since I0 contains 0x0000, we must add 0x0010 to it. Thus, 0x0010 is loaded into M0. After the first data memory read or write, which outputs 0x0000, M0 is added to the (non-bit-reversed) address in I0 so that I0 contains 0x0010. On the second data memory read or write, I0 is bit-reversed (14 bits) and the resulting address is 0x0200, the correct second bit-reversed address.

In general, the modify value is determined by raising two to the difference between 14 and the number of bits to be reversed. In this ten-bit example, the value is $2^{(14-10)} = 0x0010$. Adding this value to I0 after each memory access and reversing all 14 bits on the next memory access yields the correct bit-reversed addresses for ten bits. The first four bit-reversed addresses are shown below.

Sequence	*I0, Non-Bit-Reversed*		*I0, Bit-Reversed*	
	Hex	Binary	Hex	Binary
0	0000	00 0000 0000 0000	0000	00 0000 **0000 0000**
1	0010	00 0000 0001 0000	0200	00 00**10 0000 0000**
2	0020	00 0000 0010 0000	0100	00 00**01 0000 0000**
3	0030	00 0000 0011 0000	0300	00 00**11 0000 0000**

Only the ten least significant bits (boldface) are bit-reversed. Each time a data memory write is performed, I0 is modified by M0. Note that the modified I0 value is not bit-reversed. Bit reversal only occurs when a data memory read or write is executed.

Listings 8-8 shows the *scramble* routine which places the inputs to the DIT FFT in bit-reversed order. This module begins by initializing two constants. The first constant (*N*) is the number of input points in the FFT. The second constant (*mod_value*) is the modify value for the pointer which outputs the bit-reversed addresses. Pointers to the data buffers are initialized, and the bit-reverser is enabled for DAG1. In bit-reverse mode, any addresses output from registers I0, I1, I2, or I3 will be bit-reversed. I0 is used in *scramble*.

The scramble routine assumes real input data. In this case, the imaginary data is all zeros and can be initialized directly into the *inplaceimag* buffer. The *brev* loop consists of two instructions. First, the sequentially ordered data is read from the *input_real* buffer using I4 (from DAG2). Then, the same data is written to the bit-reversed location in the *inplacereal* buffer using I0 (from DAG1). After all the real input data has been placed in bit-reversed order in the *inplacereal* buffer, the bit-reverser is disabled for the rest of the FFT calculation.

```
{ Bit-Reverse (Scramble) Subroutine                          DIT_BREV.DSP

        Calling Parameters
            Sequentially ordered input data in inputreal

        Return Values
            Scrambled input data in inplacereal

        Altered Registers
            I0,I4,M0,M4,AY1

        Altered Memory
            inplacereal
}
.MODULE                         dit_scramble;
.CONST          N=1024,mod_value=0x0010;  {Initialize constants}
.EXTERNAL       inputreal, inplacereal;
.CONST          N=1024,mod_value=0x0010;  {Initialize constants}
.EXTERNAL       inputreal, inplacereal;
.ENTRY          scramble;

scramble:   I4=^inputreal;      {I4sequentially ordered data}
            I0=^inplacereal;    {I0scrambled data}
            M4=1;
            M0=mod_value;       {M0=modifier for reversing N bits}
            L4=0;
            L0=0;
            CNTR = N;
            ENA BIT_REV;        {Enable bit-reversed outputs on DAG1}
            DO brev UNTIL CE;
            AY1=DM(I4,M4);      {Read sequentially ordered data}
brev:       DM(I0,M0)=AY1;      {Write data in bit-reversed location}
            DIS BIT_REV;        {Disable bit-reverse}
            RTS;                {Return to calling program}

.ENDMOD;
```

—————— **Listing 8-8: Bit-Reverse (Scramble) Routine** ——————————

Block Floating-Point Scaling

Block floating-point scaling is used to maximize the dynamic range of a fixed-point data field. The block floating-point system is a hybrid between fixed-point and floating-point systems. Instead of each data word having its own exponent, the block floating-point format assumes the same exponent for an entire block of data.

The initial input data contains enough guard bits to ensure that no overflow occurs in the first FFT stage. During each stage of the FFT calculation, bit growth can occur. This bit growth can result in magnitude bits replacing guard bits. Because the stage output data is used as input data for the next stage, these guard bits must be replaced; otherwise, an output of the next stage might overflow. In a conditional block floating-point FFT, bit growth is monitored in each stage calculation. When the stage is complete, the output data of the entire stage is shifted to replace any lost guard bits.

Because a radix-2 butterfly calculation has the potential for two bits of growth, SB (the block floating-point exponent register) is initialized to -2. This sets up the ADSP-2101 block floating-point compare logic to detect any data with more than 13 bits of magnitude (or fewer than three sign bits). After each butterfly calculation, the EXPADJ instruction determines if bit growth occurred by checking the number of guard bits. For example, the value 1111 0000 0000 0000 has an exponent of -3. The value 0111 1111 1111 1111 has an exponent of zero (no guard bits). If a butterfly result has an exponent larger than the value in SB, bit growth into the guard bits has occurred, and SB is assigned the larger exponent (if it has not already been changed by bit growth in a previous butterfly of the same stage). Therefore, at the end of each stage, SB contains the exponent of the largest butterfly result(s). If no bit growth occurred, SB is not changed.

The dit_radix-2_bfp_adjust routine is shown in Listing 8-9. This routine performs block floating-point scaling on the outputs of each stage except the last of the DIT FFT. Because guard bits only need to be replaced to ensure that an output of the next stage does not overflow, the subroutine first checks to see if the block of data is the output of the last stage. If it is, no shifting is needed and the subroutine returns. If the data block is not the output of the last stage, shifting is necessary only if SB is not -2 (indicating that bit growth into guard bits occurred). If SB is -2, no bit growth occurred, so the subroutine returns.

If bit growth occurred, shifting is needed. The subroutine determines the amount to shift from the value of SB. The data can grow by either one or two bits for each stage; therefore, if bit growth occurred, SB must be either -1 or zero. If SB is -1, the data block is shifted right one bit. If SB is not -1, it must be zero. In this case, the data block is shifted right two bits. When shifting is complete, the block exponent is updated by the shifted amount (one or two).

In this routine, shifting to the right is performed through multiplication rather than shift instructions. Multiplication by an appropriate power of two gives a shifted result. For example, to shift a number two bits to the right, the number is multiplied by 0x0200. In multiplication, the product can be rounded to preserve LSB information, whereas in shifting, this information is merely lost. Multiplication thus minimizes noise.

```
{ DIT Radix-2 Block Floating-Point Scaling Routine              DIT_BFPA.DSP

    Calling Parameters
        Radix-2 DIT FFT stage results in inplacereal and inplaceimag
        Note: This code assumes inplaceimag immediately follows
              inplacereal in memory.

    Return Parameters
        inplacereal and inplaceimag adjusted for bit growth

    Altered Registers
        I0,I1,AX0,AY0,AR,MX0,MY0,MR,CNTR

    Altered Memory
        inplacereal, inplaceimag, blk_exponent
}
.MODULE         dit_radix_2_bfp_adjust;
.CONST          Ntimes2 = 2048;
.EXTERNAL       inplacereal, blk_exponent; {Begin declaration section}

.ENTRY  bfp_adj;

bfp_adj:     AY0=CNTR;                    {Check for last stage}
             AR=AY0-1
             IF EQ RTS;                   {If last stage, return}
             AY0=-2;
             AX0=SB;
             AR=AX0-AY0;                  {Check for SB=-2}
             IF EQ RTS;                   {IF SB=-2, no bit growth, return}
             I0=^inplacereal;             {I0=read pointer}
             I1=^inplacereal;             {I1=write pointer}
             AY0=-1;
             MY0=0x4000;                  {Set MY0 to shift 1 bit right}
             AR=AX0-AY0,MX0=DM(I0,M1);    {Check if SB=-1; Get first sample}
             IF EQ JUMP strt_shift;       {If SB=-1, shift block data 1 bit}
             AX0=-2;                       {Set AX0 for block exponent update}
             MY0=0x2000;                  {Set MY0 to shift 2 bits right}
strt_shift:  CNTR=Ntimes2 - 1;            {initialize loop counter}
             DO shift_loop UNTIL CE;      {Shift block of data}
                MR=MX0*MY0(RND),MX0=DM(I0,M1);
                                          {MR=shifted data,MX0=next value}
shift_loop:     DM(I1,M1)=MR1;           {Unshifted data=shifted data}
             MR=MX0*MY0(RND);             {Shift last data word}
             AY0=DM(blk_exponent);        {Update block exponent and}
             DM(I1,M1)=MR1,AR=AY0-AX0;    {store last shifted sample}
             DM(blk_exponent)=AR;
             RTS;
.ENDMOD;
```

———— **Listing 8-9: DIT Radix-2 Block Floating-Point Scaling Routine** ————

8.3.3 Exercises

1. The DIT FFT modules described in this section perform a 1024-point DFT. To implement a different size FFT requires change in the values of the constants *N* and *N_div_2* in Listing 8-1. Similarly a new set of twiddle factors *twid-real* and *twid_imag* must be computed and stored in data files. Modify the *DIT_FFT_MAIN* program (and its related modules) to compute an 8-point DFT depicted in Figure 8-2.

Determine the proper twiddle factors W_8^k and store their real and imaginary parts in data files. Create a data file containing 8 samples of the signal

$$x(n) \quad = \quad \cos\frac{\pi n}{4} \quad , \quad n = 0, 1, \ldots, 7$$

Use the Simulator to compute and verify its DFT.

2. The radix-2 DIT FFT implementation described in this section is the most compact form of the FFT, in terms of the program memory storage requirements. However, it is not the most efficient in terms of the speed of its execution. Its execution is in a fully looped form which does not exploit the unique mathematical characteristics of the first and the last stage of the FFT. Specifically, all the multiplications (equations (8-7) through (8-10)) in the first stage are by a value of either 0 or 1 and therefore can be removed. Can you write an improved DIT FFT program and its modules to incorporate the above enhancement? A solution program is available on the diskette.

8.4 RADIX-2 DECIMATION-IN-FREQUENCY FFT

In the DIT FFT, each decimation consists of two steps. First, a DFT equation is expressed as the sum of two DFTs, one of even samples and one of odd samples. This equation is then divided into two equations, one that computes the first half of the output (frequency) samples and one that computes the second half. In the decimation-in-frequency (DIF) FFT, a DFT equation is expressed as the sum of two calculations, one on the first half of the samples and one on the second half of the samples. This equation is then expressed as two equations, one that computes even output samples and one that computes odd output samples. Decimation in time refers to grouping the input sequence into even and odd samples, whereas decimation in frequency refers to grouping the output (frequency) sequence into even and odd samples. Decimation-in-frequency can thus be visualized as repeatedly dividing the output sequence into even and odd samples in the same way that decimation in time divides down the input sequence (Proakis and Manolakis, [8]).

8.4.1 The Algorithm

The DIF FFT divides an N-point DFT into two summations, shown in (8-11).

$$X(k) \quad = \quad \sum_{n=0}^{N-1} x(n) W_N^{nk}$$

$$= \quad \sum_{n=0}^{\frac{N}{2}-1} x(n) W_N^{nk} + \sum_{n=\frac{N}{2}}^{N-1} x(n) W_N^{nk}$$

$$= \quad \sum_{n=0}^{\frac{N}{2}-1} x(n) W_N^{nk} + \sum_{n=0}^{\frac{N}{2}-1} x(n+N/2) W_N^{(n+N/2)k}$$

$$(8-11)$$

Because $W_N^{(n+N/2)k} = W_N^{nk} \cdot W_N^{(N/2)k}$ and $W_N^{(N/2)k} = (-1)^k$, equation (8-11) can also be expressed as

$$X(k) \quad = \quad \sum_{n=0}^{\frac{N}{2}-1} x(n)W_N^{nk} + (-1)^k \sum_{n=0}^{\frac{N}{2}-1} x(n+N/2)W_N^{nk}$$

$$= \quad \sum_{n=0}^{\frac{N}{2}-1} [x(n) + (-1)^k x(n+N/2)]W_N^{nk} \quad , \quad k = 0,...,N-1 \qquad (8-12)$$

The decimation of the output (frequency) sequence is accomplished by dividing $X(k)$ into two equations, one that computes even output samples and one that computes odd output samples. For even values of $X(k)$, $k=2r$.

$$X(2r) \quad = \quad \sum_{n=0}^{\frac{N}{2}-1} [x(n) + (-1)^{2r} x(n+N/2)]W_N^{2nr}$$

$$= \quad \sum_{n=0}^{\frac{N}{2}-1} [x(n) + x(n+N/2)]W_{N/2}^{nr} \quad , \quad r = 0,...,\frac{N}{2}-1 \qquad (8-13)$$

For odd values of $X(k)$, $k=2r+1$.

$$X(2r+1) \quad = \quad \sum_{n=0}^{\frac{N}{2}-1} [x(n) + (-1)^{2r+1} x(n+N/2)]W_N^{(2r+1)n}$$

$$= \quad \sum_{n=0}^{\frac{N}{2}-1} [[x(n) - x(n+N/2)]W_N^n]W_{N/2}^{nr}, r = 0,...,\frac{N}{2}-1 \qquad (8-14)$$

Note that $X(2r)$ and $X(2r+1)$ are the results of $N/2$-point DFTs performed on the sum and difference of the first and second halves of the input sequence. In equation (8-14), the difference of the two halves of the input sequence is multiplied by a twiddle factor, W_N^n. Figure 8-5 illustrates the first decimation of the DIF FFT, which eliminates half $(N^2/2)$ of the DFT calculations.

Each of the two $N/2$-point DFTs $(X(2r)$ and $X(2r+1))$ are divided into two $N/4$-point DFTs in the same way as the N-point DFT is divided into two $N/2$-point DFTs. By the substitutions

$$X_1(r) \quad = \quad X(2r) \qquad\qquad r = 0,...,\frac{N}{2}-1$$

$$x_1(n) \quad = \quad x(n) + x(n+N/2) \quad n = 0,...,\frac{N}{2}-1$$

the sequence of even samples in equation (8-13) becomes

$$X_1(r) \quad = \quad \sum_{n=0}^{\frac{N}{2}-1} x_1(n)W_{N/2}^{rk} \qquad (8-15)$$

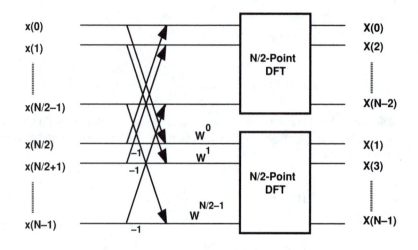

Figure 8-5: First Decimation of DIF FFT

This $N/2$-point sequence has the same form as the original N-point sequence in equation (8-11) and can be divided in half in the same manner to yield

$$X_1(r) \quad = \quad \sum_{n=0}^{\frac{N}{4}-1} [x_1(n) + (-1)^r x_1(n+N/4)] W_{N/2}^{nr} \tag{8-16}$$

For even output samples, let $r=2s$.

$$X_1(2s) \quad = \quad \sum_{n=0}^{\frac{N}{4}-1} [x_1(n) + x_1(n+N/4)] W_{N/4}^{sn} \tag{8-17}$$

For odd output samples, let $r=2s+1$.

$$X_1(2s+1) \quad = \quad \sum_{n=0}^{\frac{N}{4}-1} [(x_1(n) - x_1(n+N/4)) W_N^{2n}] W_{N/4}^{sn} \tag{8-18}$$

$X(2r+1)$ is also divided into two equations, one that computes even output samples and one that computes odd output samples, in the same way that $X(2r)$ is divided into $X_1(2s)$ and $X_1(2s+1)$. Thus we have four $N/4$-point sequences.

If we make the substitutions

$$X_2(s) \quad = \quad X_1(2s)$$
$$x_2(n) \quad = \quad x_1(n) + x_1(n+N/4)$$

equation (8-17) becomes

$$X_2(s) \quad = \quad \sum_{n=0}^{\frac{N}{4}-1} x_2(n) W_{N/4}^{sn} \qquad\qquad (8-19)$$

The four $N/4$-point sequences that result from the decimation of $X(2r)$ and $X(2r+1)$ are divided to form eight $N/8$-point sequences in the third decimation. This process is repeated until the division of a sequence results in a pair of equations that together compute a two-point DFT. In this pair, the summation variable n (see equations (8-17) and (8-18)) is equal to zero only, so no summation is performed. The two-point DFT computed by this pair of equations is the core calculation (butterfly) for the radix-2 DIF FFT.

Figure 8-6 shows the complete decimation for an eight-point DIF FFT. Notice that the inputs of the DIF FFT are in sequential order and the outputs are in scrambled order. The DIF FFT can also be performed with inputs in bit-reversed order, resulting in outputs in sequential order. In this case, however, the twiddle factors must be in bit-reversed order. In this chapter, the DIF FFT is presented with twiddle factors in sequential order to simplify programming.

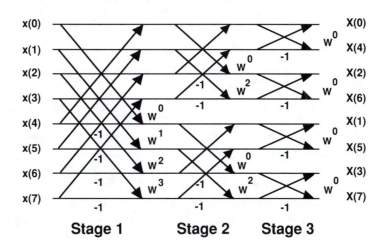

Figure 8-6: Eight-Point DIF FFT

As in the DIT FFT, the 8-point DIF FFT butterflies are organized into groups and stages. In the eight-point FFT, the first stage has one group of four ($N/2$) butterflies. The second stage has two groups of two ($N/4$) butterflies, and the last has four groups of one butterfly.

The DIF FFT butterfly is similar to that of the DIT FFT except that the twiddle factor multiplication occurs after rather than before the primary-node and dual-node subtraction. The DIF butterfly is illustrated graphically in Figure 8-7. The variables x and y represent the real and imaginary parts, respectively, of a sample. The twiddle factor can be divided into real and imaginary parts because $W_N = e^{-j2\pi/N} = \cos(2\pi/N) - j\sin(2\pi/N)$. In the program presented later in this section, the twiddle factors are initialized in memory as cosine and -sine values (not +sine). For this reason, the twiddle factors are shown in Figure 8-7 as $C + j(-S)$. C represents cosine and $-S$ represents -sine.

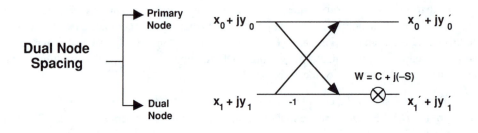

Figure 8-7: Radix-2 DIF FFT Butterfly

Equations (8-20) through (8-23) describe the DIF FFT butterfly outputs.

$$x_0' = x_0 + x_1 \tag{8-20}$$

$$y_0' = y_0 + y_1 \tag{8-21}$$

$$x_1' = C(x_0 - x_1) - (-S)(y_0 - y_1) \tag{8-22}$$

$$y_1' = (-S)(x_0 - x_1) + C(y_0 - y_1) \tag{8-23}$$

As in the DIT FFT, the butterfly is performed in-place; that is, the results of each butterfly are written over the corresponding inputs. For example, x_0' is written over x_0.

8.4.2 A DIF FFT Program

The DIF flow chart is shown in Figure 8-8. Like the DIT FFT, the DIF FFT uses three subroutines. The first subroutine computes the FFT. The second subroutine performs conditional block floating-point scaling at the end of each stage (except the last). The third subroutine bit-reverses the locations of the FFT output data to "unscramble" the data. The DIF FFT subroutine is described in this section. The block floating-point and bit reversal routines are described later in this section.

Main Module

The module *dif_fft_main* is shown in Listing 8-10. The FFT calculation is performed in one buffer (*inplacedata*). In this program, the real and imaginary input data are interleaved in the buffer. The length of *inplacedata* is thus twice the number of points in the FFT and is specified by the constant $N\_\times\_2$ ($N\_\times\_2 = 2048$ for a 1024-point FFT). Unlike the DIT FFT, the DIF FFT is performed on sequentially ordered input data and produces data in bit-reversed order; therefore, no additional buffers for scrambling the input data are needed.

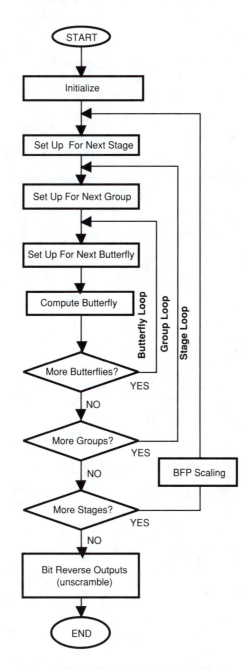

Figure 8-8: Radix-2 DIF FFT Flow Chart

When the output data is unscrambled, it is separated into real and imaginary values and placed in two buffers (*real_results*, *imaginary_results*). Twiddle-factor buffers are defined and initialized as in the DIT FFT.

The DIF FFT uses the variables *groups*, *bflys_per_group* and *blk_exponent*. Because the first stage of the DIF FFT contains one group of N/2 butterflies, *groups* is initialized to one and *bflys_per_group* is initialized to *N_div_2*. The node spacing (*node_space*) is *N* instead of *N/2* because the real and imaginary input data are interleaved.

Two subroutines are called. The first performs the DIF FFT and calls the block floating-point scaling routine. The second bit-reverses the FFT outputs to unscramble them.

```
{ DIF FFT Main program                                          DIF_MAIN.DSP

    This program computes a 1024-point DFT of data values stored in the
    buffer inplacedata using DIF FFT algorithm.  The input data is assumed
    to be complex.  The real and imaginary input data are interleaved in
    the buffer.  The input data is read from the disk file INPLACED.DAT.
    The real and imaginary parts of twiddle factors are read from disk
    files TWID_REA.DAT and TWID_IMA.DAT, respectively.  The output DFT
    values are available in data memory locations real_results and imag_
    results.
}

.MODULE/ABS=4        dif_fft_main;

.CONST               N=1024,N_x_2=2048;       {Const. for 1024 points}
.CONST               N_div_2=512,log2N=10;
.VAR/DM/RAM          inplacedata[N_x_2];
.VAR/DM/RAM          real_results[N];
.VAR/DM/RAM          imaginary_results[N];

.VAR/PM/ROM/CIRC     twid_imag[N_div_2];
.VAR/PM/ROM/CIRC     twid_real[N_div_2];

.VAR/DM/RAM          groups,node_space,bflys_per_group,blk_exponent;

.INIT                inplacedata: <inplacedata.dat>;
.INIT                twid_imag: <twid_imag.dat>;
.INIT                twid_real: <twid_real.dat>;
.INIT                groups: 1;
.INIT                node_space: N;
.INIT                bflys_per_group: N_div_2;
.INIT                blk_exponent: 0;

.GLOBAL              inplacedata, real_results, imaginary_results;
.GLOBAL              twid_real, twid_imag;
.GLOBAL              groups, bflys_per_group, node_space, blk_exponent;

.EXTERNAL            unscramble, fft_start;

                     CALL fft_start;
                     CALL unscramble;
                     TRAP;

.ENDMOD;
```
─────────── **Listing 8-10: Main Module, Radix-2 DIF FFT** ───────────

DIF FFT Module

The conditional block floating-point DIF FFT program is described in this section. The butterfly loop is described first, then the group and stage loops. The complete FFT program is presented in Listing 8-14 at the end of this section.

Butterfly Loop

The code segment for the DIF butterfly (with conditional block floating-point scaling) is shown in Listing 8-11, on the next page. The primary-node outputs x_0' and y_0' are calculated first and written over x_0 and y_0. Complex subtraction for the dual-node calculation is then performed, followed by the twiddle factor multiplication. The outputs x_1' and y_1' are written over x_1 and y_1. Instructions that write butterfly results to memory are boldface. Each butterfly output is checked for bit growth using the EXPADJ instruction. This loop is repeated *bflys_per_group* times.

The input and output parameters for the butterfly loop are as follows:

Initial Conditions	*Final Conditions*
$AX0 = x_0$	$AX0 = $ next x_0
$AY0 = x_1$	$AY0 = $ next x_1
$AY1 = y_1$	$AY1 = $ next y_1
$I0 \rightarrow y_0$	$I0 \rightarrow $ next y_0
$I1 \rightarrow $ next x_1	$I1 \rightarrow x_1$ after next
$I2 \rightarrow x_1$	$I2 \rightarrow $ next x_1
$I4 \rightarrow C$	$CNTR = $ butterfly count -1
$I5 \rightarrow (-S)$	
$M0 = -1$	
$M1 = 1$	
$M5 = $ twiddle factor modifier	
$CNTR = $ butterfly count	

```
AR=AX0+AY0,AX1=DM(I0,M0),MY0=PM(I4,M5);   {AR=x0+x1,AX1=y0,MY0=C,I0 x0}
SB=EXPADJ AR;                             {Check for bit growth}
DM(I0,M1)=AR,AR=AX1+AY1;                   {x0´=x0+x1,AR=y0+y1,I0 y0}
SB=EXPADJ AR;                             {Check for bit growth}
DM(I0,M1)=AR,AR=AX0-AY0;                   {y0´=y0+y1,AR=x0-x1,I0 next x0}
MX0=AR,AR=AX1-AY1;                         {MX0=x0-x1,AR=y0-y1}
MR=MX0*MY0(SS),AX0=DM(I0,M1),MY1=PM(I5,M5);
                                          {MR=(x0-x1)C,AX0=next x0,MY1=(-S)}
MR=MR-AR*MY1(RND),AY0=DM(I1,M1);          {MR=(x0-x1)C-(y0-y1)(-S),}
                                          { AY0=next x1}
SB=EXPADJ MR1;                            {Check for bit growth}
DM(I2,M1)=MR1,MR=AR*MY0 (SS);             {x1´=(x0-x1)C-(y0-y1)(-S),}
                                          { MR=(y0-y1)C}
MR=MR+MX0*MY1(RND),AY1=DM(I1,M1);         {MR=(y0-y1)C+(x0-x1)(-S),}
                                          { AY1=next y1}
DM(I2,M1)=MR1,SB=EXPADJ MR1;              {y1´=(y0-y1)C+(x0-x1)(-S),check}
                                          {for bit growth}
```

———————— Listing 8-11: Radix-2 DIF FFT Butterfly, Conditional Block Floating-Point ————

Group Loop

The group loop code is shown in Listing 8-12. The group loop sets up the butterfly loop by fetching initial data and initializing the butterfly loop counter. When all the butterflies in a group have been calculated, data pointers are updated to point to the inputs for the first butterfly of the next group. This loop is repeated until all groups in a stage are complete. The input and output parameters of the group loop are as follows:

Initial Conditions   Final Conditions

$I0 \rightarrow x_0$ of first butterfly in group $I0 \rightarrow x_0$ of first butterfly in next group
$I1 \rightarrow x_1$ of first butterfly in group $I1 \rightarrow x_1$ of first butterfly in next group
$I2 \rightarrow x_1$ of first butterfly in group $I2 \rightarrow x_1$ of first butterfly in next group
CNTR = group count   CNTR = group count -1
M1 = 1
M2 = *node_space*
M3 = *node_space-2*

```
              CNTR=DM(bflys_per_group);      {Initialize butterfly counter}
              AX0=DM(I0,M1);               {AX0=x0}
              AY0=DM(I1,M1);               {AY0=x1}
              AY1=DM(I1,M1);               {AY1=y1}
              DO bfly_loop UNTIL CE;
bfly_loop:        {Calculate All Butterflies in Group}
              MODIFY(I2,M2);            {I2 ->x1 of 1st butterfly in next group}
              MODIFY(I1,M3);            {I1 ->x1 of 1st butterfly in next group}
              MODIFY(I0,M3);
group_loop:MODIFY(I0,M1);              {I0 ->x0 of 1st butterfly in next group}
```
—————————— **Listing 8-12: Radix-2 DIF FFT Group Loop** ——————————

Stage Loop

The stage loop code is shown in Listing 8-13. This code segment sets up and computes all groups in a stage and controls stage characteristics, such as the number of groups in a stage. Pointers I0 and I1 are set to point to x_0 and x_1 of the first butterfly in the first group of the stage. Pointer I2 also points to x_1 and is used to write the dual-node butterfly results to data memory. M3 is set to *node_space-2* and is used to modify pointers for the next group. The group counter is initialized to groups, the number of groups in the stage. The twiddle factor modifier stored in M5 is also *groups*. This value is the exponent increment value for the twiddle factors of consecutive butterflies in a group.

The SB register is set to -2 to detect any bit growth into the guard bits of any butterfly output. When all the groups in a stage are computed, the *bfp_adjustment* routine is called to check for bit growth and adjust the output data if necessary. Then parameters for the next stage are updated; *groups* is doubled and *node_space* and *bflys_per_group* are divided in half. The stage loop is repeated $\log_2 N$ times.

The input and output parameters of the stage loop are summarized below. Note that the parameters are passed in memory locations.

Initial Conditions   *Final Conditions*

groups = # groups in stage   *groups* = # groups in next stage
bflys_per_group = # butterflies/group   *bflys_per_group* = #butterflies/
                                                      group (next stage)

node_space = node spacing current stage   *node_space* = node spacing next
                                                      stage

inplacedata = stage input data   *inplacedata* = stage output data

```
                I0=^inplacedata;        {I0 ->x0 in 1st butterfly of stage}
                I1=^inplacedata;
                AY0=DM(node_space);
                M2=AY0;                 {M2=dual node spacing}
                MODIFY(I1,M2);          {I1 ->x1 in 1st butterfly of stage}
                I2=I1;
                AX0=2;
                AR=AY0-AX0;
                M3=AR;                  {M3=node_space-2}
                CNTR=DM(groups);        {Initialize group counter}
                SB=-2;                  {Set minimum allowable sign bits to two}
                M5=DM(groups);          {M5=twiddle factor modifier}
                DO group_loop UNTIL CE;

group_loop:             {Calculate All Groups in Stage}

                CALL bfp_adjustment;    {Adjust block data for bit growth}
                SI=DM(groups);
                SR=LSHIFT SI BY 1 (LO);
                DM(groups)=SR0;         {groups=groups × 2}
                SI=DM(node_space);
                SR=LSHIFT SI BY -1 (LO);
                DM(node_space)=SR0;     {node_space=node_space ÷ 2}
                SR=LSHIFT SR0 BY -1(LO);
                DM(bflys_per_group)=SR0; {bflys_per_group=bflys_per_group ÷ 2}
```

—————————————— **Listing 8-13: Radix-2 DIF FFT Stage Loop** ——————————————

DIF FFT Subroutine

The complete block floating-point DIF FFT subroutine is shown in Listing 8-14. Initializations of index, modifier and length registers that retain the same values throughout the FFT calculation are performed before the stage loop is entered. Instructions that write butterfly results to memory are boldface.

```
{ Radix-2 DIF FFT Subroutine                                           R2DIFFFT.DSP

    Performs Radix-2 DIF FFT

    Calling Parameters
            inplacedata = Real input data in sequential order
            twid_real = Twiddle factor cosine values
            twid_imag = Twiddle factor sine values
            groups = N/2
            bflys_per_group = 1
            node_space = 1

    Return Values
            inplacedata = Real and Imaginary FFT results interleaved in
                          sequential order

.MODULE   dif_fft;

.CONST    N=1024, N_div_2=512, logN=10;

.EXTERNAL inplacedata, twid_real, twid_imag;
.EXTERNAL groups,bflys_per_group,node_space;
.EXTERNAL bfp_adjust;

.ENTRY    fft_start;

fft_start:  I4=^twid_real;  {I4 -> C OF W⁰}
            L4=N_div_2;
            I5=^twid_imag;   {I5 -> (-S) OF W⁰}
```

```
              L5=N_div_2;
              M0=-1;
              M1=1;
              CNTR=logN;                {Initialize stage counter}
              L0=0;
              L1=0;
              L2=0;
              DO stage_loop UNTIL CE;
                 I0=^inplacedata;          {I0 -> x0}
                 I1=^inplacedata;
                 AY0=DM(node_space);
                 M2=AY0;
                 MODIFY(I1,M2);         {I1 -> x1}
                 I2=I1;
                 AX0=2;
                 AR=AY0-AX0;
                 M3=AR;                       {M3=node_space-2}
                 CNTR=DM(groups);             {Initialize group counter}
                 SB=-2;
                 M5=CNTR;              {Init. twiddle factor modifier}
                 DO group_loop UNTIL CE;
                    CNTR=DM(bflys_per_group);    {Init. butterfly counter}
                    AX0=DM(I0,M1);               {AX0=x0}
                    AY0=DM(I1,M1);               {AY0=x1}
                    AY1=DM(I1,M1);               {AY1=y1}
                    DO bfly_loop UNTIL CE;
                       AR=AX0+AY0, AX1=DM(I0,M0), MY0=PM(I4,M5);
                                          {AR=x0+x1,AX1=y0,MY0=C}
                       SB=EXPADJ AR;              {Check for bit growth}
                       DM(I0,M1)=AR,AR=AX1+AY1;     {x0´=x0+x1,AR=y0+y1}
                       SB=EXPADJ AR;              {Check for bit growth}
                       DM(I0,M1)=AR,AR=AX0-AY0;     {y0´=y0+y1,AR=x0-x1}
                       MX0=AR, AR=AX1-AY1;  {MX0=x0-x1,AR=y0-y1}
                       MR=MX0*MY0 (SS), AX0=DM(I0,M1), MY1=PM(I5,M5);
                                          {MR=(x0-x1)C,AX0=next x0,MY1=(-S)}
                       MR=MR-AR*MY1 (RND), AY0=DM(I1,M1);
                                          {MR=(x0-x1)C-(y0-y1)(-S),AY0=next x1}
                       SB=EXPADJ MR1;            {Check for bit growth}
                       DM(I2,M1)=MR1, MR=AR*MY0 (SS);
                                          {x1´=(x0-x1)C-(y0-y1)(-S),MR=(y0-y1)C}
                       MR=MR+MX0*MY1 (RND), AY1=DM(I1,M1);
                                          {MR=(y0-y1)C+(x0-x1)(-S),AY1=next y1}
bfly_loop:                 DM(I2,M1)=MR1, SB=EXPADJ MR1;
                                          {y1´=(y0-y1)C+(x0-x1)(-S),check bit growth}
                    MODIFY(I2,M2);   {I2->x1 of first butterfly in next group}
                    MODIFY(I1,M3);   {I1->x1 of first butterfly in next group}
                    MODIFY(I0,M3);
group_loop:         MODIFY(I0,M1);   {I0->x0 of first butterfly in next group}
                 CALL bfp_adjust;    {Adjust block data for bit growth}
                 SI=DM(groups);
                 SR=LSHIFT SI BY 1 (LO);
                 DM(groups)=SR0;      {groups=groups × 2}
                 SI=DM(node_space);
                 SR=LSHIFT SI BY -1 (LO);
                 DM(node_space)=SR0;             {node_space=node_space + 2}
                 SR=LSHIFT SR0 BY -1 (LO);
stage_loop:      DM(bflys_per_group)=SR0;
                                  {bflys_per_group=bflys_per_group + 2}

              RTS;

.ENDMOD;
```

—————— **Listing 8-14: Radix-2 DIF FFT Routine, Conditional Block Floating-Point** ——————

Bit Reversal

As described earlier, the output sequence of the radix-2 DIF FFT is in a bit reversed order which must be unscrambled after the FFT is performed. This unscrambling is also accomplished through bit reversal. The basic scheme and code of bit reversal algorithm is described in Section 8.3

Listings 8-15 shows the *unscramble* routine which places the output data of the DIF FFT in sequential order. The module begins by initializing two constants. The first constant (*N*) is the number of input points in the FFT. The second constant (*mod_value*) is the modify value for the pointer which outputs the bit-reversed addresses. Pointers to the data buffers are initialized, and the bit-reverser is enabled for DAG1. In bit-reverse mode, any addresses output from registers I0, I1, I2, or I3 will be bit-reversed. The I1 register is used in *unscramble*.

The *unscramble* routine uses two loops: one to unscramble the real FFT output data, the other to unscramble the imaginary output data. I4 points to the first of the scrambled real data values in the *inplacedata* buffer. I4 is modified by two (in M4) after each read. Because the real and imaginary data in *inplacedata* are interleaved, this ensures that only real data is read for the first loop. I1 contains the (bit-reversed) address of the first location in the *real_results* buffer (for unscrambled real data). The appropriate modify value (stored in M1) is added to I1 upon each data memory write. Before entering the second loop, I0 is updated to point to the first imaginary data in *inplacedata* and I1 is set to the first address (bit-reversed) of the *imag_results* buffer (for sequentially ordered imaginary data).

```
{ Bit-Reverse (Uncramble) Subroutine                DIF_BREV.DSP

      Calling Parameters
          Real and imaginary scrambled output data in inplacedata

      Return Values
          Sequentially ordered real output data in real_results
          Sequentially ordered imag. output data in imaginary_results

      Altered Registers
          I0,I1,I4,M1,M4, AY1,CNTR

      Altered Memory
          real_results, imaginary_results
}
.MODULE       dif_unscramble;
.CONST        N=1024,mod_value=0x0010;  {Initialize constants}
.EXTERNAL     inplacedata;
.ENTRY        unscramble;                        {Declare entry point into module}

unscramble:   I4=^inplacedata;       {I4real part of 1st data point}
              M4=2;                  {Modify by 2 to fetch only real data}
              L0=0;
              L4=0;
              I1=0x4;                {I1=1st real output addr, bit-reversed}
              M1=mod_value;          {Modifier for 10-bit reversal}
              CNTR=N;                {N=number of real data points}
              ENA BIT_REV;           {Enable bit-reverse}
              DO bit_rev_real UNTIL CE;
                  AY1=DM(I4,M4);     {Read real data}
```

```
bit_rev_real:   DM(I1,M1)=AY1;       {Place in sequential order}
                I4=^inplacedata+1;   {I4imag. part of 1st data point}
                I1=0xC;              {I1=1st imag. output addr,bit-reversed}
                CNTR=N;              {N=number of imaginary data points}
                DO bit_rev_imag UNTIL CE;
                    AY1=DM(I4,M4);   {Read imag. data}
bit_rev_imag:   DM(I1,M1)=AY1;       {Place in sequential order}
                DIS BIT_REV;         {Disable bit-reverse}
                RTS;

.ENDMOD;
```

———————————— **Listing 8-15: Bit-Reverse (Unscramble) Routine** ————————————

Block Floating-Point Scaling

In Section 8.3, we discussed the block floating-point scaling algorithm used in the DIT FFT subroutine. The scaling is used to maximize the dynamic range of a fixed-point data field. This block floating-point routine, *dit_radix-2_bfp_adjust*, can be modified for the DIF FFT routine by replacing *inplacereal* references with *inplacedata*. The modified routine, *dif_radix-2_bfp_adjust*, is shown in Listing 8-16. It performs block floating-point scaling on the outputs of each stage except the last of the DIF FFT.

```
{ DIF Radix-2 Block Floating-Point Scaling Routine          DIF_BFPA.DSP

      Calling Parameters
          Radix-2 DIF FFT stage results in inplacedata

      Return Parameters
          inplacedata adjusted for bit growth

      Altered Registers
          I0,I1,AX0,AY0,AR,MX0,MY0,MR,CNTR

      Altered Memory
          inplacedata, blk_exponent
}

.MODULE     dif_radix_2_bfp_adjust;
.CONST      Ntimes2 = 2048;
.EXTERNAL   inplacedata, blk_exponent; {Begin declaration section}

.ENTRY   bfp_adjust;

bfp_adjust: AY0=CNTR;                    {Check for last stage}
            AR=AY0-1
            IF EQ RTS;                   {If last stage, return}
            AY0=-2;
            AX0=SB;
            AR=AX0-AY0;                  {Check for SB=-2}
            IF EQ RTS;                   {IF SB=-2, no bit growth, return}
            I0=^inplacedata;             {I0=read pointer}
            I1=^inplacedata;             {I1=write pointer}
            AY0=-1;
            MY0=0x4000;                  {Set MY0 to shift 1 bit right}
            AR=AX0-AY0,MX0=DM(I0,M1);    {Check if SB=-1; Get first sample}
            IF EQ JUMP strt_shift;       {If SB=-1, shift block data 1 bit}
            AX0=-2;                      {Set AX0 for block exponent update}
            MY0=0x2000;                  {Set MY0 to shift 2 bits right}
strt_shift: CNTR=Ntimes2 - 1;           {initialize loop counter}
            DO shift_loop UNTIL CE;      {Shift block of data}
              MR=MX0*MY0(RND),MX0=DM(I0,M1);
                                         {MR=shifted data,MX0=next value}
shift_loop:   DM(I1,M1)=MR1;            {Unshifted data=shifted data}
```

```
    MR=MX0*MY0(RND);            {Shift last data word}
    AY0=DM(blk_exponent);       {Update block exponent and}
    DM(I1,M1)=MR1,AR=AY0-AX0;   {store last shifted sample}
    DM(blk_exponent)=AR;
    RTS;
.ENDMOD;
```
─────────── **Listing 8-16: DIF Radix-2 Block Floating-Point Scaling Routine** ───────────

8.4.3 Exercises

1. The DIF FFT modules described in this section perform a 1024-point DFT. To implement a different size FFT requires change in the values of the constants N, N_x_2, N_div_2, and $logN$ in Listing 8-10. Similarly a new set of twiddle factors *twid-real* and *twid_imag* must be computed and stored in data files. Modify the *DIF_FFT_MAIN* program (and related modules) to compute an 8-point DFT depicted in Figure 8-6. Determine the proper twiddle factors W_8^k and store their real and imaginary parts in data files. Create a data file containing 8 samples of the signal

$$x(n) \;=\; \cos\frac{\pi n}{4} \quad , \quad n = 0, 1, ..., 7$$

    Use the Simulator to compute and verify its DFT.

2. A radix-2 FFT devides an N-point sequence successively in half until only two-point DFTs remain. Similarly, a radix-4 FFT devides an N-point sequence succesively in quarters until only four-point DFTs remain. The four-point DFT is the core calculation (butterfly) of the radix-4 FFT. Refer to Proakis and Manolakis [8] for more details and the algorithm. Write a main program, a DIF FFT module *rad4_fft*, and a digit reversal module *digit_rev* to implement a 16-point DFT. A solution is available on the diskette.

8.5 THE INVERSE DFT AND THE IFFT ALGORITHM

The inverse relationship for obtaining a sequence from its DFT is called the inverse DFT (IDFT). It is given by the equation

$$x(n) \;=\; \sum_{k=0}^{N-1} X(k)W_n^{-nk} \quad , \quad n = 0, ..., N-1 \tag{8-24}$$

Although the FFT algorithms described in the last two sections were presented in the context of computing the DFT efficiently, they may also be used in computing the IDFT.

The only difference between the two transforms is the normalization factor $1/N$ and the phase sign of the twiddle factor W_N. Consequently, an FFT algorithm for computing the DFT may be converted into an IFFT algorithm for computing the IDFT by using a reversed (upside down) twiddle factor table and by dividing the output of the algorithm by N.

8.6 SUMMARY

The implementation of the DFT on the ADSP-2101 using fast algorithms was the topic of this chapter. In particular, we developed the decimation-in-time and decimation-in-frequency fast Fourier transform algorithms and described detailed procedures to implement these algorithms in assembly language. These by no means are the only efficient algorithms available to compute the DFT. Several other excellent algorithms are available in the literature. One such algorithm is the Radix-4 FFT which is widely used in practice. The discussion provided in this chapter should help the user develop programs for this and other algorithms. Furthermore, the programs given in this chapter can be used as building blocks for developing programs in spectrum analysis, correlation analysis, and frequency domain linear filtering.

The remaining two chapters describe applications of digital signal processing in digital communications and adaptive filtering using programs and routines developed so far.

chapter 9

APPLICATIONS IN COMMUNICATIONS

9.1 INTRODUCTION

Today, microprocessors and microcomputers find widespread use in the implementation of a variety of electronic systems. In this chapter we shall focus on several applications dealing with waveform representation and coding, especially speech coding, and with digital communications. In particular, we shall describe several methods for digitizing analog waveforms, with specific application to speech coding and transmission. These methods are pulse-code modulation (PCM), differential PCM and adaptive differential PCM (ADPCM) delta modulation (DM) and adaptive delta modulation (ADM), and linear predictive coding (LPC). An experiment is formulated involving each of these waveform encoding methods for implementation on the ADSP-2101 family of microcomputers.

The last three topics treated in this chapter deal with signal detection applications that are usually encountered in the implementation of a receiver in a digital communication system. For each of these topics we describe an experiment that involves the implementation of the detection scheme on an ADSP-2101 microcomputer.

9.2 PULSE CODE MODULATION

Pulse code modulation is a method for quantizing an analog signal for the purpose of transmitting or storing the signal in digital form. PCM is widely used for speech transmission in telephone communications and for telemetry systems that employ radio transmission. We shall concentrate our attention on the application of PCM to speech signal processing.

Speech signals transmitted over telephone channels are usually limited in bandwidth to the frequency range below 4kHz. Hence, the Nyquist rate for sampling such a signal is less than 8kHz. In PCM, the analog speech signal is sampled at the nominal rate of 8kHz (samples per second) and each sample is quantized to one of 2^b levels, and represented digitally by a sequence of b bits. Thus, the bit rate required to transmit the digitized speech signal is 8000.b bits per second.

The quantization process may be modeled mathematically as

$$\tilde{s}(n) \;=\; s(n)+q(n) \tag{9-1}$$

where $\tilde{s}(n)$ represents the quantized value of $s(n)$ and $q(n)$ represents the quantization error which we treat as an additive noise. Assuming that a uniform quantizer is used and the number of levels is sufficiently large, the quantization noise is well characterized statistically by the uniform probability density function,

$$p(q) \;=\; \frac{1}{\Delta}, \quad -\frac{\Delta}{2}\le q \le \frac{\Delta}{2} \tag{9-2}$$

where the step size of the quantizer is $\Delta = 2^{-b}$. The mean square value of the quantization error is

$$E(q^2) \;=\; \frac{\Delta^2}{12} \;=\; \frac{2^{-2b}}{12} \tag{9-3}$$

Measured in decibels, the mean square value of the noise is

$$10\log\!\left(\frac{\Delta^2}{12}\right) \;=\; 10\log\!\left(\frac{2^{-2b}}{12}\right) \;=\; -6b-10.8\text{dB} \tag{9-4}$$

We observe that the quantization noise decreases by 6 dB/bit used in the quantizer. High quality speech requires a minimum of 12 bits per sample and, hence, a bit rate of 96,000 bits per second (bps).

Speech signals have the characteristic that small signal amplitudes occur more frequently than large signal amplitudes. However, a uniform quantizer provides the same spacing between successive levels throughout the entire dynamic range of the signal. A better approach is to use a nonuniform quantizer which provides more closely spaced levels at the low signal amplitudes and more widely spaced levels at the large signal amplitudes. For a nonuniform quantizer with b bits, the resulting quantization error has a mean square value that is smaller than that given by (9-4). A nonuniform quantizer characteristic is usually obtained by passing the signal through a nonlinear device that compresses the signal amplitude, followed by a uniform quantizer. For example, a logarithmic compressor employed in U.S. and Canadian telecommunications systems has an input-output magnitude characteristic of the form

$$|y| \;=\; \frac{\log(1+\mu|s|)}{\log(1+\mu)} \tag{9-5}$$

where $|s|$ is the magnitude of the input, $|y|$ is the magnitude of the output, and μ is a parameter that is selected to give the desired compression characteristic.

In the encoding of speech waveforms the value of $\mu = 255$ has been adopted as a standard in the U.S. and Canada. This value results in about a 24dB reduction in the quantization noise power relative to uniform quantization. Consequently, an 8-bit quantizer used in conjunction with a $\mu = 255$ logarithmic compressor produces the same quality speech as a 12-bit uniform quantizer with no compression. Thus, the compressed PCM speech signal, has a bit rate of 64,000 bps.

The logarithmic compressor standard used in European telecommunication systems is called A-law and is defined as

$$|y| = \frac{1 + \log(A|s|)}{1 + \log A} \qquad (9-6)$$

where A is chosen as 87.56. Although (9-5) and (9-6) are different nonlinear functions, the two compression characteristics are very similar. Figure 9-1 illustrates these two compression functions. We note that they are very similar.

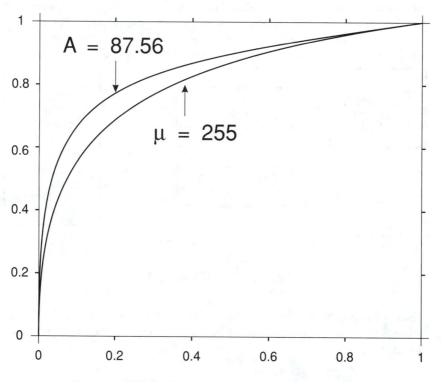

Figure 9-1: Comparison of μ-law and A-law Nonlinearities

In the reconstruction of the signal from the quantized values, the decoder employs an inverse logarithmic relation to expand the signal amplitude. The combined compressor-expandor pair is termed a *compandor*.

In an implementation of the logarithmic compressor, the logarithmic function is approximated by a piecewise linear function. Eight straight line segments along the curve result in a close approximation to the logarithmic function. A sample value of the signal is represented by its sign (positive or negative), the straight-line segment on the logarithmic approximation (three bits to specify the one of eight segments), and the position of the sample on the particular line segment. Thus, the μ-law PCM 8-bit word consists of the following three parts: (1) the most significant bit (MSB) is the sign bit; (2) the next three bits represent the straight line segment number; (3) the last four bits represent the position within the segment.

PCM Experiment

The purpose of this experiment is to gain an understanding of the PCM compression (linear-to-logarithmic) and the PCM expansion (logarithmic-to- linear) algorithm.

Use the two software modules, one for $\mu = 255$ compression and for $\mu = 255$ expansion to study the effect of logarithmic companding. Create data files of different waveforms and pass them through the μ-law compressor and expander. Display the input and output waveforms as indicated in Figure 9-2, and comment on the results. Include an exponential waveform and a saw-tooth waveform among the input signals. A waveform generator may also be substituted for the waveform data file. If the microphone input is used for the input port, modify your program to use only PCM expansion.

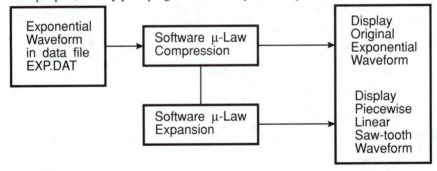

Figure 9-2: PCM Experiment

Note that since the compressed PCM code is an 8-bit word, it is shifted (left justified) and displayed. Also note that decoding uses full fractional numbers (1.15 format) rather than the integer format.

9.3 DIFFERENTIAL PCM (DPCM) AND ADAPTIVE DPCM (ADPCM)

In PCM each sample of the waveform is encoded independently of all the other samples. However, most signals including speech sampled at the Nyquist rate or faster exhibit significant correlation between successive samples. In other words, the average change in amplitude between successive samples is relatively small. Consequently an encoding scheme that exploits the redundancy in the samples will result in a lower bit rate for the speech signal.

9.3.1 Differential PCM

A relatively simple solution is to encode the differences between successive samples rather than the samples themselves. Since differences between samples are expected to be smaller than the actual sampled amplitudes, fewer bits are required to represent the differences. A refinement of this general approach is to predict the current sample based on the previous p samples. To be specific, let $s(n)$ denote the current sample of speech and let $\hat{s}(n)$ denote the predicted value of $s(n)$ defined as

$$\hat{s}(n) \;=\; \sum_{i=1}^{p} a(i)s(n-i) \tag{9-7}$$

Thus $\hat{s}(n)$ is a weighted linear combination of the past p samples and the $a(i)$ are the predictor (filter) coefficients. The $a(i)$ are selected to minimize some function of the error between $s(n)$ and $\hat{s}(n)$.

A mathematically and practically convenient error function is the sum of squared errors. With this as the performance index for the predictor, we select the $a(i)$ to minimize

$$\mathcal{E}_p \;=\; \sum_{n=1}^{N} e^2(n) \;=\; \sum_{n=1}^{N} \left[s(n) - \sum_{i=1}^{p} a(i)s(n-i) \right]^2$$

$$\mathcal{E}_p \;=\; r_{ss}(0) - 2\sum_{i=1}^{p} a(i)r_{ss}(i) + \sum_{i=1}^{p}\sum_{j=1}^{p} a(i)a(j)r_{ss}(i-j) \tag{9-8}$$

where $r_{ss}(m)$ is the autocorrelation function of the sampled signal sequence $s(n)$ defined as

$$r_{ss}(m) \;=\; \sum_{i=1}^{N} s(i)s(i+m) \tag{9-9}$$

Minimization of $\mathcal{E}_p$ with respect to the predictor coefficient $a(i)$ results in the set of linear equations, called the normal equations

$$\sum_{i=1}^{p} a(i)r_{ss}(i-j) \;=\; r_{ss}(j), \quad j=1,2,\ldots,p \tag{9-10}$$

Thus the values of the predictor coefficients are established.

Having described the method for determining the predictor coefficients, let us now consider the block diagram of a practical DPCM system, shown in Figure 9-3. In this configuration, the predictor is implemented with the feedback loop around the quantizer. The input to the predictor is denoted as $\tilde{s}(n)$ which represents the signal sample $s(n)$ modified by the quantization process, and the output of the predictor is

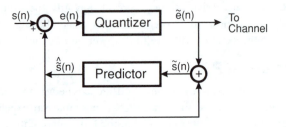

Figure 9-3: (a) Block Diagram of a DPCM Encoder

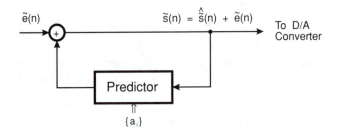

Figure 9-3: (b) DPCM Decoder at the Receiver

$$\hat{s}(n) \;=\; \sum_{i=1}^{p} a(i)s(n-i) \tag{9-11}$$

The difference

$$e(n) \;=\; s(n)-\hat{s}(n) \tag{9-12}$$

is the input to the quantizer and $\tilde{e}(n)$ denotes the output. Each value of the quantized prediction error $\tilde{e}(n)$ is encoded into a sequence of binary digits and transmitted over the channel to the receiver. The quantized error $\tilde{e}(n)$ is also added to the predicted value $\hat{\tilde{s}}(n)$ to yield $\tilde{s}(n)$.

At the receiver the same predictor that was used at the transmitting end is synthesized and its output $\hat{\tilde{s}}(n)$ is added to $\tilde{e}(n)$ to yield $\tilde{s}(n)$. The signal $\tilde{s}(n)$ is the desired excitation for the predictor and also the desired output sequence from which the reconstructed signal $\tilde{s}(t)$ is obtained by filtering, as shown in Figure 9-3b.

The use of feedback around the quantizer, as described above, ensures that the error in $\tilde{s}(n)$ is simply the quantization error $q(n) = \tilde{e}(n) - e(n)$ and that there is no accumulation of previous quantization errors in the implementation of the decoder. That is,

$$\begin{aligned} q(n) \;&=\; \tilde{e}(n)-e(n) \\ &=\; \tilde{e}(n)-s(n)-\tilde{s}(n) \\ &=\; \tilde{s}(n)-s(n) \end{aligned} \tag{9-13}$$

Hence $\tilde{s}(n) = s(n) + q(n)$. This means that the quantized sample $\tilde{s}(n)$ differs from the input $s(n)$ by the quantization error $q(n)$ independent of the predictor used. Therefore the quantization errors do not accumulate.

In the DPCM system illustrated in Figure 9-3, the estimate or predicted value $\tilde{s}(n)$ of the signal sample $s(n)$ is obtained by taking a linear combination of past values $\tilde{s}(n-k), k = 1, 2, \ldots, p$, as indicated by (9-11). An improvement in the quality of the estimate is obtained by including linearly filtered past values of the quantized error. Specifically, the $\tilde{s}(n)$ estimate may be expressed as

$$\hat{\tilde{s}}(n) \;=\; \sum_{i=1}^{p} a(i)\tilde{s}(n-i) + \sum_{i=1}^{m} b(i)\tilde{e}(n-i) \qquad (9-14)$$

where $b(i)$ are the coefficients of the filter for the quantized error sequence $\tilde{e}(n)$. The block diagram of the encoder at the transmitter and the decoder at the receiver are shown in Figure 9-4. The two sets of coefficients $a(i)$ and $b(i)$ are selected to minimize some function of the error $e(n) = \tilde{s}(n) - s(n)$ such as the sum of squared errors.

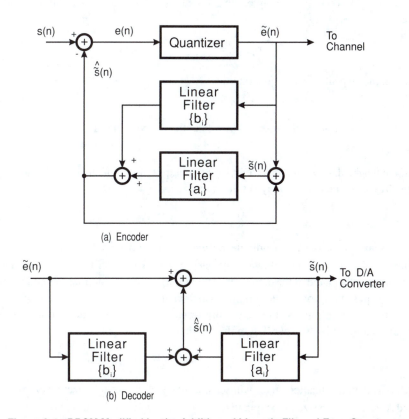

(a) Encoder

(b) Decoder

Figure 9-4: DPCM Modified by the Addition of Linearly Filtered Error Sequence

By using a logarithmic compressor and a 4-bit quantizer for the error sequence $e(n)$, DPCM results in high quality speech at a rate of 32,000 bps, which is a factor of two lower than PCM.

9.3.2 Adaptive PCM and DPCM

In general, the power in a speech signal varies slowly with time. PCM and DPCM encoders, however, are designed on the basis that the speech signal power is constant and, hence, the quantizer is fixed. The efficiency and performance of these encoders can be improved by having them adapt to the slowly time-variant power level of the speech signal.

In both PCM and DPCM, the quantization error $q(n)$ resulting from a uniform quantizer operating on a slowly varying power level input signal will have a time-variant variance (quantization noise power). One improvement which reduces the dynamic range of the quantization noise is the use of an adaptive quantizer.

Adaptive quantizers can be classified as feedforward or feedback. A feedforward adaptive quantizer adjusts its step size for each signal sample based on a measurement of the input speech signal variance (power). For example, the estimated variance based as a sliding window estimator is

$$\hat{\sigma}_{n+1}^2 = \frac{1}{M} \sum_{k=n+1-M}^{n+1} s^2(k) \tag{9-15}$$

Then, the step size for the quantizer is

$$\Delta(n+1) = \Delta(n)\hat{\sigma}_{n+1} \tag{9-16}$$

In this case, it is necessary to transmit $\Delta(n+1)$ to the decoder in order for it to reconstruct the signal.

A feedback adaptive quantizer employs the output of the quantizer in the adjustment of the step size. In particular, we may set the step size as

$$\Delta(n+1) = \alpha(n)\Delta(n) \tag{9-17}$$

where the scale factor $\alpha(n)$ depends on the previous quantizer output. For example, if the previous quantizer output is small, we may select $\alpha(n) < 1$ in order to provide for finer quantization. On the other hand, if the quantizer output is large, then the step size should be increased to reduce the possibility of signal clipping. Such an algorithm has been successfully used in the encoding of speech signals. Figure 9-5 illustrates such a (3- bit) quantizer in which the step size is adjusted recursively according to the relation

$$\Delta(n+1) = \Delta(n) \cdot M(n)$$

where $M(n)$ is a multiplication factor whose value depends on the quantizer level for the sample $s(n)$ and $\Delta(n)$ is the step size of the quantizer for processing $s(n)$. Values of the multiplications factors optimized for speech encoding have been given by Jayant [18]. These values are displayed in Table 9-1 for 2-, 3-, and 4-bit quantization, for PCM and DPCM.

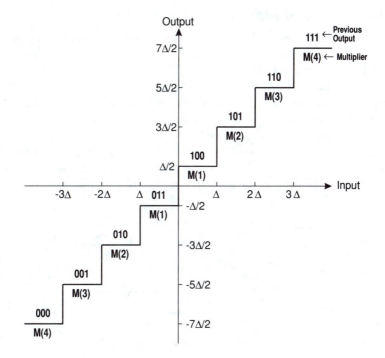

Figure 9-5: Example of a Quantizer with an Adaptive Step Size (Jayant [18])

	PCM			DPCM		
	2	3	4	2	3	4
$M(1)$	0.60	0.85	0.80	0.80	0.90	0.90
$M(2)$	2.20	1.00	0.80	1.60	0.90	0.90
$M(3)$		1.00	0.80		1.25	0.90
$M(4)$		1.50	0.80		1.70	0.90
$M(5)$			1.20			1.20
$M(6)$			1.60			1.60
$M(7)$			2.00			2.00
$M(8)$			2.40			2.40

Table 9-1: Multiplication Factors for Adaptive Step Size Adjustment (Jayant [18]) ——————

In DPCM, the predictor can also be made adaptive. Thus, in ADPCM the coefficients of the predictor are changed periodically to reflect the changing signal statistics of the speech. The linear equations given by (9-10) still apply, but the short-term autocorrelation function of $s(n)$, $r_{ss}(m)$ changes with time.

9.3.3 ADPCM Standard

Figure 9-6 illustrates, in block diagram form, a 32,000 bps ADPCM encoder and decoder that has been adopted as an international (CCITT) standard for speech transmission over telephone channels. The ADPCM encoder is designed to accept 8-bit PCM compressed signal samples at 64,000 bps and by means of adaptive prediction and adaptive 4-bit quantization reduces the bit rate over the channel to 32,000 bps. The ADPCM decoder accepts the 32,000 bps data stream and reconstructs the signal in the form of an 8-bit compressed PCM at 64,000 bps. Thus, we have a configuration shown in Figure 9-7, where the ADPCM encoder/decoder is embedded into a PCM system. Although the ADPCM encoder/decoder could be used directly on the speech signal, the interface to the PCM system is necessary in practice in order to maintain compatibility with existing PCM systems that are widely used in the telephone network.

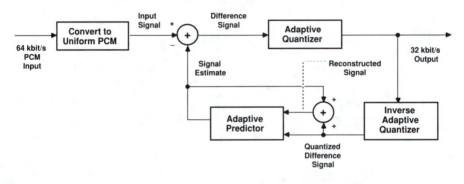

ENCODER

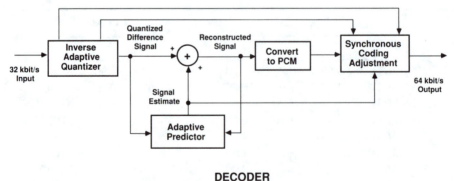

DECODER

Figure 9-6: ADPCM Block Diagram

The ADPCM encoder accepts the 8-bit PCM compressed signal and expands it to a 14-bit per sample linear representation for processing. The predicted value is subtracted from this

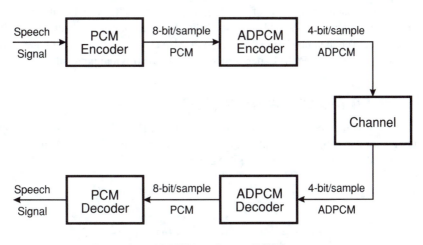

Figure 9-7: ADPCM Interface to PCM System

14-bit linear value to produce a difference signal sample that is fed to the quantizer. Adaptive quantization is performed on the difference signal to produce a 4-bit output for transmission over the channel.

Both the encoder and decoder update their internal variables based only on the ADPCM values that are generated. Consequently, an ADPCM decoder including an inverse adaptive quantizer is embedded in the encoder so that all internal variables are updated based on the same data. This ensures that the encoder and decoder operate on synchronism without the need to transmit any information on the values of internal variables.

The adaptive predictor computes a weighted average of the last six dequantized difference values and the last two predicted values. Hence, this predictor is basically a two-pole ($p = 2$) and six-zero ($m = 6$) filter governed by the difference equation given by (9-14). The filter coefficients are updated adaptively for every new input sample.

At the receiving decoder and at the decoder that is embedded in the encoder, the 4-bit transmitted ADPCM value is used to update the inverse adaptive quantizer whose output is a dequantized version of the difference signal. This dequantized value is added to the value generated by the adaptive predictor to produce the reconstructed speech sample. This signal is the output of the decoder which is converted to compressed PCM format at the receiver.

ADPCM Experiment

The objective of this experiment is to gain familiarity and understanding of ADPCM and its interface with a PCM encoder/decoder (transcoder). As described above the ADPCM transcoder is inserted between the PCM compressor and the PCM expander, as shown in Figure 9-7. This is the configuration of the software PCM and DPCM modules for this experiment.

The input to the PCM-ADPCM transcoder system can be supplied either from internally generated waveform data files, or from a microphone input, or from an external waveform generator, just as in the case of the PCM experiment. The output of the transcoder can be monitored and displayed from one of the DAC output ports. Comparisons should be made between the output signal from the PCM-ADPCM transcoder with the signal from the PCM trancoder (PCM Experiment), and with the original input signal.

9.4 DELTA MODULATION (DM)

Delta modulation may be viewed as a simplified form of DPCM in which a two- level (1-bit) quantizer is used in conjunction with a fixed first-order predictor. The block diagram of a DM encoder-decoder is shown in Figure 9-8a. We note that

$$\hat{\tilde{s}}(n) \;=\; \tilde{s}(n-1) \;=\; \hat{\tilde{s}}(n-1)+\tilde{e}(n-1) \tag{9-18}$$

Since

$$q(n) \;=\; \tilde{e}(n)-e(n) \;=\; \tilde{e}(n)-[s(n)-\hat{\tilde{s}}(n)]$$

It follows that

$$\hat{\tilde{s}}(n) \;=\; s(n-1)+q(n-1) \tag{9-19}$$

Thus the estimated (predicted) value of $s(n)$ is really the previous sample $s(n-1)$ modified by the quantization noise $q(n-1)$. We also note that the difference equation in (9-18) represents an integrator with an input $\tilde{e}(n)$. Hence an equivalent realization of the one-step predictor is an accumulator with an input equal to the quantized error signal $\tilde{e}(n)$. In general, the quantized error signal is scaled by some value, say Δ_1, which is called the *step size*. This equivalent realization is illustrated in Figure 9-8b. In effect, the encoder shown in Figure 9-8b approximates a waveform $s(t)$ by a linear staircase function. In order for the approximation to be relatively good, the waveform $s(t)$ must change slowly relative to the sampling rate. This requirement implies that the sampling rate must be several (a factor of at least 5) times the Nyquist rate. A lowpass filter is usually incorporated into the decoder to smooth out discontinuities in the reconstructed signal.

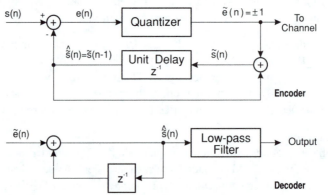

Figure 9-8: (a) Block Diagram of a Delta Modulation System

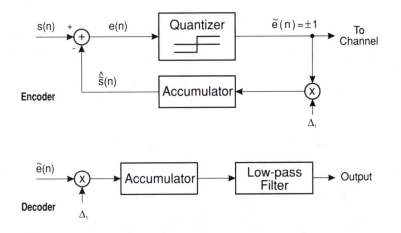

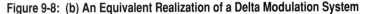

Figure 9-8: (b) An Equivalent Realization of a Delta Modulation System

9.4.1 Adaptive Delta Modulation (ADM)

At any given sampling rate, the performance of the DM encoder is limited by the two types of distortion. One is called *slope-overload distortion*. It is due to the use of a step size Δ_1 that is too small to follow portions of the waveform that have a steep slope. The second type of distortion, called *granular noise*, results from using a step size that is too large in parts of the waveform having a small slope. The need to minimize both of these two types of distortion results in conflicting requirements in the selection of the step size Δ_1.

An alternative solution is to employ a variable size that adapts itself to the short-term characteristics of the source signal. That is, the step size is increased when the waveform has a steep slope and decreased when the waveform has a relatively small slope.

A variety of methods can be used to set adaptively the step size in every iteration. The quantized error sequence $\tilde{e}(n)$ provides a good indication of the slope characteristics of the waveform being encoded. When the quantized error $\tilde{e}(n)$ is changing signs between successive iterations, this is an indication that the slope of the waveform in the locality is relatively small. On the other hand, when the waveform has a steep slope, successive values of the error $\tilde{e}(n)$ are expected to have identical signs. From these observations it is possible to devise algorithms which decrease or increase the step size depending on successive values of $\tilde{e}(n)$. A relatively simple rule devised by Jayant [17] is to vary adaptively the step size according to the relation

$$\Delta(n) \;=\; \Delta(n-1)K^{\tilde{e}(n)\tilde{e}(n-1)}, \quad n=1,2,\ldots \qquad (9-20)$$

where $K \geq 1$ is a constant that is selected to minimize the total distortion. A block diagram of a DM encoder-decoder that incorporates this adaptive algorithm is illustrated in Figure 9-9.

Several other variations of adaptive DM encoding have been investigated and described in the technical literature. A particularly effective and popular technique first proposed by Greefkes [15] is called *continuously variable slope delta modulation* (CVSD). In CVSD the adaptive step-size parameter may be expressed as

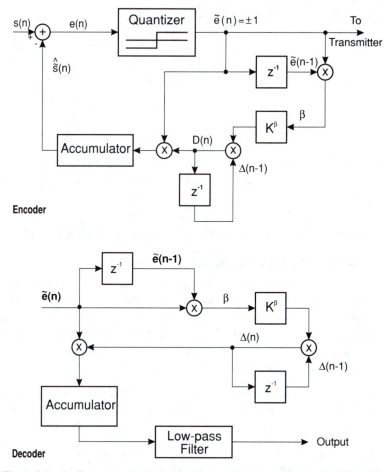

Figure 9-9: An Example of a Delta Modulation System with Adaptive Step Size

$$\Delta(n) \quad = \quad \alpha\Delta(n-1)+k_1 \qquad\qquad (9-21)$$

if $\tilde{e}(n)$, $\tilde{e}(n-1)$, and $\tilde{e}(n-2)$ have the same sign; otherwise

$$\Delta(n) \quad = \quad \alpha\Delta(n-1)+k_2 \qquad\qquad (9-22)$$

The parameters α, k_1, and k_2 are selected such that $0 < \alpha < 1$ and $k_1 > k_2 > 0$. For more discussion on this and other variations of adaptive DM, the interested reader is referred to the papers by Jayant [18] and Flanagan et al. [13] and to the extensive references contained in these papers.

DM and ADM Experiment

The purpose of this experiment is to gain an understanding of delta modulation and adaptive delta modulation for coding of waveforms. This experiment involves writing software modules for the DM encoder and decoder as shown in Figure 9-8, and for the

ADM encoder and decoder shown in Figure 9-9. The lowpass filter at the decoder can be implemented as a linear-phase FIR filter. For example, a Hanning filter which has the impulse response

$$h(n) \; = \; \frac{1}{2}\left(1 - \cos\frac{2\pi n}{N-1}\right), \quad 0 \le n \le N-1 \tag{9.23}$$

may be used, where the length N may be selected in the range $5 \le N \le 15$.

The input to the DM and ADM systems can be supplied either from internally generated waveform data files, or from an external waveform generator, or from a microphone input, just as in the case of the PCM Experiment. The output of the decoder can be monitored and displayed from one of the DAC output ports. Comparisons should be made between the output signal from the DM and ADM decoders and the original input signal.

9.5 LINEAR PREDICTIVE CODING (LPC) OF SPEECH

The linear predictive coding (LPC) method for speech analysis and synthesis is based on modeling the vocal tract as a linear all-pole (IIR) filter having the system function

$$H(z) \; = \; \frac{G}{1 + \displaystyle\sum_{k=1}^{p} a_p(k)z^{-k}} \tag{9-24}$$

where p is the number of poles, G is the filter gain, and $\{a_p(k)\}$ are the parameters that determine the poles. There are two mutually exclusive excitation functions to model voiced and unvoiced speech sounds. On a short-time basis, voiced speech is periodic with a fundamental frequency F_0, or a pitch period $1/F_0$, which depends on the speaker. Thus voiced speech is generated by exciting the all-pole filter model by a periodic impulse train with a period equal to the desired pitch period. Unvoiced speech sounds are generated by exciting the all-pole filter model by the output of a random- noise generator. This model is shown in Figure 9-10.

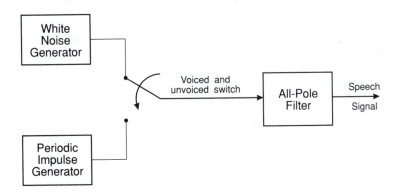

Figure 9-10: Block Diagram Model for the Generation of a Speech Signal

Given a short-time segment of a speech signal, usually about 20 ms or 160 samples at an 8kHz sampling rate, the speech encoder at the transmitter must determine the proper excitation function, the pitch period for voiced speech, the gain parameter G, and the coefficients $a_p(k)$. A block diagram that illustrates the speech encoding system is given in Figure 9-11. The parameters of the model are determined adaptively from the data and encoded into a binary sequence and transmitted to the receiver. At the receiver, the speech signal is synthesized from the model and the excitation signal.

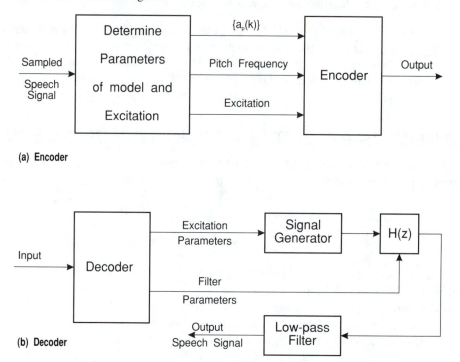

Figure 9-11: Encoder and Decoder for LPC

The parameters of the all-pole filter model are easily determined from the speech samples by means of linear prediction. To be specific, the output of the FIR filter is

$$\hat{s}(n) \;=\; -\sum_{k=1}^{p} a_p(k)s(n-k) \qquad (9-25)$$

and the corresponding error between the observed sample $s(n)$ and the estimate $\hat{s}(n)$ is

$$e(n) \;=\; s(n) + \sum_{k=1}^{p} a_p(k)s(n-k) \qquad (9-26)$$

By minimizing the sum of squared errors, i.e.

$$\mathcal{E} = \sum_{n=0}^{N} e^2(n) = \sum_{n=0}^{N} \left[s(n) + \sum_{k=1}^{p} a_p(k)s(n-k) \right]^2 \qquad (9-27)$$

we can determine the pole parameters $\{a_p(k)\}$ of the model. The result of differentiating $\mathcal{E}$ with respect to each of the parameters and equating the result to zero, is a set of p linear equations

$$\sum_{k=1}^{p} a_p(k)r_{ss}(m-k) = -r_{ss}(m), \quad m = 1, 2, ..., p \qquad (9-28)$$

where $r_{ss}(m)$ is the autocorrelation of the sequence $s(n)$ defined as

$$r_{ss}(m) = \sum_{n=0}^{N} s(n)s(n+m) \qquad (9-29)$$

These equations can be solved recursively and most efficiently, without resorting to matrix inversion, by using the Levinson-Durbin algorithm (for reference see Levinson [20], Durbin [12], or Proakis and Manolakis [8]). These recursive equations are

$$a_m(m) = K_m = -\frac{r_{ss}(m) + \sum_{k=1}^{m-1} a_{m-1}(k)r_{ss}(m-k)}{\mathcal{E}_{m-1}}, m = 1, 2, ..., p$$

$$a_m(k) = a_{m-1}(k) + K_m a_{m-1}(m-k) \qquad (9-30)$$

$$= a_{m-1}(k) + a_m(m)a_{m-1}(m-k) \quad \begin{matrix} k = 1, 2, ..., m-1 \\ m = 1, 2, ..., p \end{matrix}$$

$$\mathcal{E}_m = \mathcal{E}_{m-1}(1 - K_m^2) \quad , \quad m = 1, 2, ..., p \quad ; \quad \mathcal{E}_0 = r_{ss}(0)$$

where $\{K_m\}$ are the reflection coefficients in the equivalent lattice filter. The prediction coefficients in the all pole model are $\{a_p(k)\}$ and the residual prediction squared error is $\mathcal{E}_p$.

The gain parameter of the filter can be obtained by noting that its input- output equation is

$$s(n) = -\sum_{k=1}^{p} a_p(k)s(n-k) + Gx(n) \qquad (9-31)$$

where $x(n)$ is the input sequence. Clearly,

$$Gx(n) = s(n) + \sum_{k=1}^{p} a_p(k)s(n-k) = e(n)$$

Then

$$G^2 \sum_{n=0}^{N-1} x^2(n) = \sum_{n=0}^{N-1} e^2(n) \qquad (9-32)$$

If the input excitation is normalized to unit energy be design, then

$$G^2 \;=\; \sum_{n=0}^{N-1} e^2(n) \;=\; r_{ss}(0) + \sum_{k=1}^{p} a_p(k) r_{ss}(k) \qquad (9.33)$$

Thus, G^2 is set equal to the residual energy resulting from the least squares optimization.

Once the LPC coefficients are computed, we can determine whether the input speech frame is voiced, and if so, what the pitch is. This is accomplished by computing the sequence

$$r_e(n) \;=\; \sum_{k=1}^{p} r_a(k) r_{ss}(n-k) \qquad (9-34)$$

where $r_a(k)$ is defined as

$$r_a(k) \;=\; \sum_{i=1}^{p} a_p(i) a_p(i+k) \qquad (9-35)$$

which is the autocorrelation sequence of the prediction coefficients. The pitch is detected by finding the peak of the normalized sequence $r_e(n)/r_e(0)$ in the time interval that corresponds to 3 to 15 ms in the 20 ms sampling frame. If the value of this peak is at least 0.25, the frame of speech is considered voiced with a pitch period equal to the value of $n = N_p$, where $r_e(N_p)/r_e(0)$ is a maximum. If the peak value is less than 0.25, the frame of speech is considered unvoiced and the pitch is zero.

The values of the LPC coefficients, the pitch period and the type of excitation are transmitted to the receiver where the decoder synthesizes the speech signal by passing the proper excitation through the all-pole filter model of the vocal tract. Typically, the pitch period requires 6 bits and the gain parameter may be represented by 5 bits after its dynamic range is compressed logarithmically. If the prediction coefficients were to be coded, they would require between 8 to 10 bits per coefficients for accurate representation. The reason for such high accuracy is that relatively small changes in the prediction coefficients result in a large change in the pole positions of the filter model. The accuracy requirements are lessened by transmitting the reflection coefficients, $\{K_i\}$ which have a smaller dynamic range, i.e. $|K_i| \le 1$. These are adequately represented by 6 bits per coefficient. Thus, for a 10th-order predictor the total number of bits assigned to the model parameters per frame is 72. If the model parameters are changed every 20 milliseconds, the resulting bit rate is 3,600 bps. Since the reflection coefficients are usually transmitted to the receiver, the synthesis filter at the receiver is implemented as an all-pole lattice filter, described in Chapter 7.

LPC Experiment

The objective of this experiment is to synthesize a speech signal that has been processed through an LPC coder. The decoder that performs the synthesis is an all-pole lattice whose parameters are the reflection coefficients that have been pre-computed by the LPC speech analyzer. The additional parameters required are the gain G, the type for excitation, and if the excitation is a periodic impulse train for voiced speech, we also need the pitch period. The output of this experiment is a speech signal that can be compared with the original speech signal. The distortion effects due to LPC analysis/synthesis may be assessed qualitatively.

9.6 DUAL-TONE MULTIFREQUENCY (DTMF) SIGNALS

DTMF is the generic name for push-button telephone signaling that is equivalent to the Touch Tone system in use within the Bell System. DTMF also finds widespread use in electronic mail systems and telephone banking systems in which the user can select options from a menu by sending DTMF signals from a telephone.

In a DTMF signaling system a combination of a high frequency tone and a low frequency tone represent a specific digit or the characters * and #. There are eight frequencies which are arranged as shown in Figure 9-12, to accommodate a total of 16 characters, 12 of which are assigned as shown while the other four are reserved for future use.

		Column 1 1209Hz	Column 2 1336Hz	Column 3 1477Hz	Column 4 1633Hz
Row 1	697Hz	1	2	3	A
Row 2	770Hz	4	5	6	B
Row 3	852Hz	7	8	9	C
Row 4	941Hz	*	0	#	D

DTMF digit = Row Tone + Column Tone

Figure 9-12: DTMF Digits

DTMF signals are easily generated in software on a microcomputer and detected by means of digital filters, also implemented in software, that are tuned to the eight frequency tones. Usually, DTMF signals are interfaced to the analog world via a codec (coder/decoder) chip or by linear A/D and D/A converters. Codec chips contain all the necessary A/D and D/A, sampling, and filtering circuitry for a bi-directional analog/digital interface.

The ADSP-2101 has been programmed to read DTMF digits stored in data memory in a relocatable look-up list. Alternatively, a DTMF keypad could be used for digit entry. In either case, the resultant DTMF tones may be generated either mathematically or from a look-up table. In the ADSP-2101, digital samples of two sine waves are generated mathematically, scaled, and added together. The sum is logarithmically compressed and sent to the codec for conversion to an analog signal. At an 8kHz sampling rate, the ADSP-2101 must output a sample every

125 ms. In this case, a sine look-up table is not used because the values of the sinewave can be computed quickly without using the large amount of data memory that a table look-up would require.

At the receiving end, the ADSP-2101 reads the logarithmically compressed, 8- bit digital data words from the codec, logarithmically expands each 8-bit sample to its 16-bit linear format and then detects the tones to decide on the transmitted digit. The detection algorithm can be a DFT implementation using the FFT algorithm or a filter bank implementation. For the relatively small number of tones to be detected, the filter bank implementation is more efficient. Below, we describe the use of the Goertzel algorithm to implement the eight tuned filters.

Recall from the discussion in Chapter 8 that the DFT of an N-point data sequence $\{x(n)\}$ is

$$X(k) = \sum_{n=0}^{N-1} x(n)W_N^{nk} \quad , \quad k = 0, 1, \ldots, N-1 \qquad (9-36)$$

If the FFT algorithm is used to perform the computation of the DFT, the number of computations (complex multiplications and additions) is $N \log_2 N$. In this case, we obtain all N values of the DFT at once. However, if we desire to compute only M points of the DFT, where $M < \log_2 N$, then a direct computation of the DFT is more efficient. The Goertzel algorithm, which is described below, is basically a linear filtering approach to the computation of the DFT, and provides an alternative to direct computation.

9.6.1 The Goertzel Algorithm

The Goertzel algorithm exploits the periodicity of the phase factors $\{W_N^k\}$ and allows us to express the computation of the DFT as a linear filtering operation. Since $W_N^{-kN} = 1$, we can multiply the DFT by this factor. Thus

$$X(k) = W_N^{-kN} \sum_{m=0}^{N-1} x(m)W_N^{-k(N-m)} \qquad (9-37)$$

We note that (9-37) is in the form of a convolution. Indeed, if we define the sequence $y_k(n)$ as

$$y_k(n) = \sum_{m=0}^{N-1} x(m)W_N^{-k(n-m)} \qquad (9-38)$$

then it is clear that $y_k(n)$ is the convolution of the finite-duration input sequence $x(n)$ of length N with a filter that has an impulse response

$$h_k(n) = W_N^{-kn}u(n) \qquad (9-39)$$

The output of this filter at $n = N$ yields the value of the DFT at the frequency $\omega_k = 2\pi k/N$. That is,

$$X(k) = y_k(n)\big|_{n=N} \qquad (9-40)$$

as can be verified by comparing (9-37) with (9-38).

The filter with impulse response $h_k(n)$ has the system function

$$H_k(z) \;=\; \frac{1}{1 - W_N^{-k} z^{-1}} \qquad\qquad (9-41)$$

This filter has a pole on the unit circle at the frequency $\omega_k = 2\pi k/N$. Thus the entire DFT can be computed by passing the block of input data into a parallel bank of N single-pole filters (resonators), where each filter has a pole at the corresponding frequency of the DFT.

Instead of performing the computation of the DFT as in (9-38), via convolution, we can use the difference equation corresponding to the filter given by (9-41) to compute $y_k(n)$ recursively. Thus we have

$$y_k(n) \;=\; W_N^{-k} \, y_k(n-1) + x(n) \quad , \quad y_k(-1) = 0 \qquad\qquad (9-42)$$

The desired output is $X(k) = y_k(N)$. To perform this computation, we can compute once and store the phase factor W_N^{-k}.

The complex multiplications and additions inherent in (9-42) can be avoided by combining the pairs of resonators possessing complex-conjugate poles. This leads to two-pole filters with system functions of the form

$$H_k(z) \;=\; \frac{1 - W_N^{k} z^{-1}}{1 - 2\cos(\omega\pi k/N) z^{-1} + z^{-2}} \qquad\qquad (9-43)$$

The realization of the system illustrated in Figure 9-13 is described by the difference equations

$$v_k(n) \;=\; 2\cos\frac{2\pi k}{N} v_k(n-1) - v_k(n-2) + x(n) \qquad\qquad (9-44)$$

$$y_k(n) \;=\; v_k(n) - W_N^{k} v_k(n-1) \qquad\qquad (9-45)$$

with initial conditions $v_k(-1) = v_k(-2) = 0$. This is the Goertzel algorithm.

The recursive relations in (9-44) is iterated for $n = 0, 1, \ldots, N$, but the equation in (9-45) is computed only once at a time $n = N$. Each iteration requires one real multiplication and two additions. Consequently, for a real input sequence $x(n)$, this algorithm requires $N+1$ real multiplications to yield not only $X(k)$ but also, due to symmetry, the value of $X(N-k)$.

We can now implement the DTMF decoder by use of the Goertzel algorithm. Since there are eight possible tones to be detected, we require eight filters of the type given by (9-43), with each filter tuned to one of the eight frequencies. In the DTMF detector, there is no need to compute the complex value $X(k)$; only the magnitude $|X(k)|$ or the magnitude square value $|X(k)|^2$ will suffice. Consequently, the final step in the computation of the DFT value involving the numerator term (feedforward part of the filter computation) can be simplified. In particular, we have

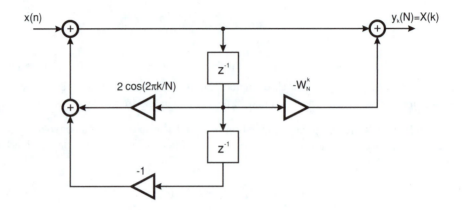

x(n)   y_k(N)=X(k)

$2\cos(2\pi k/N)$

$-W_N^k$

Figure 9-13: Realization of Two-Pole Resonator for Computing the DFT

$$|X(k)|^2 \;=\; |y_k(N)|^2 \;=\; |v_k(N) - W_N^k v_k(N-1)|^2$$

$$=\; v_k^2(N) + v_k^2(N-1) - \left(2\cos\frac{2\pi k}{N}\right) v_k(N)v_k(N-1) \qquad (9-46)$$

Thus, complex-valued arithmetic operations are completely eliminated in the DTMF detector.

DTMF Experiment

The objective of this experiment is to gain an understanding of the DTMF tone generation software and the DTMF decoding algorithm (the Goertzel algorithm). In this experiment, a dialing sequence of several digits may be stored in data memory, DIAL-LIST. The DTMF digits are stored in the four least significant bits of the 16-bit data word as follows:

DTMF Digit	Hex Word	DTMF Digit	Hex Word
1	0x0000	7	0x8000
2	0x1000	8	0x9000
3	0x2000	9	0xA000
A	0x3000	C	0xB000
4	0x4000	*	0xC000
5	0x5000	0	0xD000
6	0x6000	#	0xE000
B	0x7000	D	0xF000

Delimiter used in the dial list;

Stop dialing	0xFxxx
Redial	0x0Fxx
Quiet space	0x00Fx

A simple finite state machine is implemented in which the IRQ2 switch is used to advance the state. The states are as follows:

State 0	No digits are generated
State 1	A continuous dial tone (350 Hz + 440 Hz) is generated
State 2	DTMF Dialing Occurs

When a Stop Dialing Delimiter is read in the DIAL-LIST, the machine jumps back to State 0.

A single-channel DTMF decoder has been implemented in software for this experiment. The valid decoded DTMF digits are sent to the DAC port for display. The relationship between the received DTMF digits and the hexadecimal code output is as follows:

DTMF Digit	Output Code	DTMF Digit	Output Code
1	0x0000	7	0x8000
2	0x1000	8	0x9000
3	0x2000	9	0xA000
A	0x3000	C	0xB000
4	0x4000	*	0xC000
5	0x5000	0	0xD000
6	0x6000	#	0xE000
B	0x7000	D	0xF000
Invalid	0xFFFF (bad_digit_code)		

In addition to exercising the program described above, we suggest that the student use a spectrum analyzer to observe the frequency components of the generated tones.

9.7 BINARY DIGITAL COMMUNICATIONS

Digitized speech signals that have been encoded via PCM, ADPCM, DM and LPC are usually transmitted to the decoder by means of digital modulation. A binary digital communications system employs two signal waveforms, say $s_1(t) = s(t)$ and $s_2(t) = -s(t)$ to transmit the binary sequence representing the speech signal. The signal waveform $s(t)$, which is nonzero over the interval $0 \leq t \leq T$, is transmitted to the receiver if the data bit is a 1 and the signal waveform $-s(t)$, $0 \leq t \leq T$ is transmitted if the data bit is a 0. The time interval T is called the signal interval and the bit rate over the channel is $R = 1/T$ bits per second. A typical signal waveform $s(t)$ is a rectangular pulse, i.e., $s(t) = A$, $0 \leq t \leq T$, which has energy A^2T.

In practice, the signal waveforms transmitted over the channel are corrupted by additive noise and other types of channel distortions that ultimately limit the performance of the communications system. As a measure of performance we normally use the average probability of error, which is often called the bit error rate.

Experiment on Binary Data Communications System

The purpose of this experiment is to investigate the performance of a binary data communications system on an additive noise channel by means of simulation. The basic configuration of the system to be simulated is shown in Fig. 9-14. Five software modules are required.

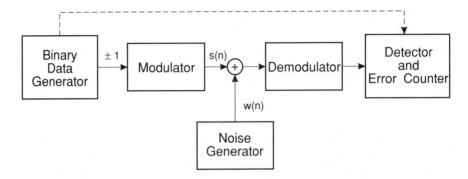

Figure 9-14: Model of Binary Data Communications System

1. A binary data generator module that generates a sequence of independent binary digits with equal probability.

2. A modulator module that maps a binary digit 1 into a sequence of M consecutive +1's, and maps a binary digit 0 into a sequence of M consecutive -1's. Thus, the M consecutive +1's represent a sampled version of the rectangular pulse.

3. A noise generator that generates a sequence of uniformly distributed numbers over the interval *(-a, a)*. Each noise sample is added to a corresponding signal sample.

4. A demodulator module that sums the M successive outputs of the noise corrupted sequence + 1's or -1's received from the channel. We assume that the demodulator is time synchronized so that it knows the beginning and end of each waveform.

5. A detector and error counting module. The detector compares the output of the modulator with zero and decides in favor of 1 if the output is greater than zero and in favor of 0 if the output is less than zero. If the output of the detector does not agree with the transmitted bit from the transmitter, an error is counted by the counter. The error rate depends on the ratio of the size of M to the additive noise power, which is $P_n = a^2/12$.

A Flowchart of the simulation program is illustrated in Fig. 9-15. The measured error rate can be displayed for different signal-to-noise ratios, either by changing M and keeping P_n fixed or vice versa.

9.8 SPREAD SPECTRUM COMMUNICATIONS

Spread spectrum signals are often used in the transmission of digital data over communication channels that are corrupted by interference due to intentional jamming or from other users of the channel. In applications other than communications, spread spectrum signals are used to obtain accurate range (time delay) and range rate (velocity) measurements in radar and navigation. For the sake of brevity, we shall limit our discussion to the use of spread spectrum for digital communications. Such signals have the characteristic that their bandwidth is much greater than the information rate in bits per second.

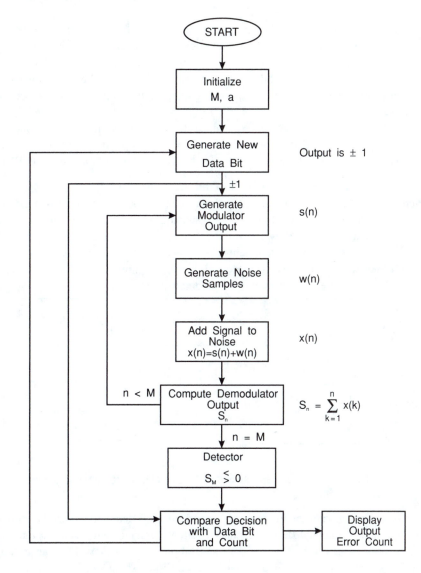

Figure 9-15: Flowchart for Simulating a Binary Data Communications System

In combatting intentional interference (jamming), it is important to the communicators that the jammer who is trying to disrupt their communication does not have prior knowledge of the signal characteristics. To accomplish this, the transmitter introduces an element of unpredictability or randomness (pseudo-randomness) in each of the possible transmitted signal waveforms, which is known to the intended receiver, but not to the jammer. As a consequence, the jammer must transmit an interfering signal without knowledge of the pseudo-random characteristics of the desired signal.

Interference from other users arises in multiple-access communications systems in which the number of users share a common communications channel. At any given time, a subset of these users may transmit information simultaneously over a common channel to corresponding receivers. The transmitted signals in this common channel may be distinguished from one another by superimposing a different pseudo-random pattern, called a *multiple-access code*, in each transmitted signal. Thus, a particular receiver can recover the transmitted data intended for it by knowing the pseudo-random pattern, i.e., the key used by the corresponding transmitter. This type of communication technique, which allows multiple users to simultaneously use a common channel for data transmission is called *code division multiple access* (CDMA).

The block diagram shown in Fig. 9-16 illustrates the basic elements of a spread spectrum digital communications system. It differs from a conventional digital communications system by the inclusion of two identical pseudo-random pattern generators, one which interfaces with the modulator at the transmitting end and the second which interfaces with the demodulator at the receiving end. The generators generate a pseudo-random or *pseudo-noise* (PN) binary-valued sequence (± 1's) which is impressed on the transmitted signal at the modulator and removed from the received signal at the demodulator.

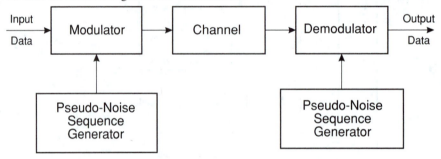

Figure 9-16: Basic Spread Spectrum Digital Communications System

Synchronization of the PN sequence generated at the demodulator with the PN sequence contained in the incoming received signal is required in order to demodulate the received signal. Initially, prior to the transmission of data, synchronization is achieved by transmitting a short fixed PN sequence to the receiver for purposes of establishing synchronization. After time synchronization of the PN generators is established, the transmission of data commences.

Experiment on Binary Spread Spectrum Communications

The objective of this experiment is to demonstrate the effectiveness of a PN spread spectrum signal in suppressing sinusoidal interference. Let us consider the binary communication system described in the experiment of Section 9-7, and let us multiply the output of the modulator by a binary (±1) PN sequence. The same binary PN sequence is used to multiply the input to the demodulator and, thus, to remove the effect of the PN sequence in the desired signal. The channel corrupts the transmitted signal by the addition of a wideband noise sequence {w(n)} and a sinusoidal interference sequence of the form $i(n) = A \sin \omega_0 n$, where $0 < \omega_0 < \pi$. We may assume that $A \geq M$, where M is the number of samples per bit from the modulator. The basic binary spread spectrum system is shown in Fig. 9-17. As can be observed, this is just the binary digital communication system shown in Fig. 9-14, to which we have added the sinusoidal interference and the PN sequence generators.

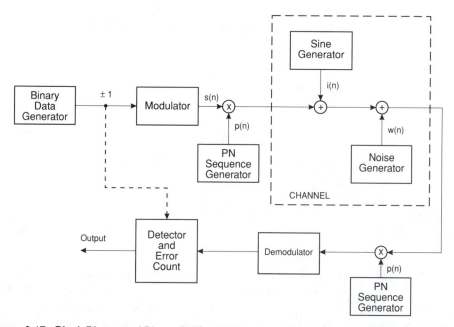

Figure 9-17: Block Diagram of Binary PN Spread Spectrum System for Simulation Experiment

The PN sequence may be generated by using a random number generator to generate a sequence of equal-probable ± 1's.

Draw a flowchart and simulate the spread-spectrum system shown in Fig. 9-17. Run the simulated system with and without the use of the PN sequence and measure the error rate under the condition that $A \geq M$ for different values of M, e.g., $M = 50, 100, 500, 1000$. Explain the effect of the PN sequence on the sinusoidal interference signal. Thus explain why the PN spread spectrum system outperforms the conventional binary communication system in the presence of the sinusoidal jamming signal.

9.9 SUMMARY

In the chapter we focused on applications of the ADSP-2101 to waveform representation and coding. In particular, we described several methods for digitizing analog waveform, including PCM, DPCM, ADPCM, DM, ADM, and LPC. These methods have been widely used for speech coding and transmission. Experiments involving these waveform encoding methods were formulated for implementation on the ADSP-2100 family of microcomputers.

We also described signal detection and communication systems where the ADSP-2101 may be used to perform the signal processing tasks. Experiments were also devised for these applications.

chapter 10

ADAPTIVE FILTERS
AND THEIR APPLICATIONS

10.1 INTRODUCTION

In Chapters 6 and 7 we described methods for designing FIR and IIR digital filters to satisfy some desired specifications. Our goal was to determine the coefficients of the digital filter that met the desired specifications.

In contrast to the filter design techniques considered in these two chapters, there are many digital signal processing applications in which the filter coefficients cannot be specified a priori. For example, let us consider a high-speed modem that is designed to transmit data over telephone channels. Such a modem employs a channel equalizer to compensate for the channel distortion. The modem must effectively transmit data through communication channels that have different frequency response characteristics and hence result in different distortion effects. The only way in which this is possible is if the channel equalizer has *adjustable coefficients* that can be optimized to minimize some measure of the distortion, on the basis of measurements performed on the characteristics of the channel. Such a filter with adjustable parameters is called an *adaptive filter*, in this case an *adaptive equalizer*.

Numerous applications of adaptive filters have been described in the literature. Some of the more noteworthy applications include (1) adaptive antenna systems in which adaptive filters are used for beam steering and for providing nulls in the beam pattern to remove undesired interference (for a reference, see the paper by Widrow et al. [22]); (2) digital communication receivers in which adaptive filters are used to provide equalization of intersymbol interference and for channel identification (for reference see Proakis [21]); (3) adaptive noise canceling techniques in which an adaptive filter is used to estimate and eliminate a noise component in some desired signal (for reference, see papers by Widrow et al. [23], Hsu and Giordano [16], and Ketchum and Proakis [19]); (4) system modeling, in which an adaptive filter is used as a model to estimate the characteristics of an unknown system. These are just a few of the best known examples on the use of adaptive filters.

274

Although both IIR and FIR filters have been considered for adaptive filtering, the FIR filter is by far the most practical and widely used. The reason for this preference is quite simple. The FIR filter has only adjustable zeros and hence it is free of stability problems associated with adaptive IIR filters that have adjustable poles as well as zeros. We should not conclude, however, that adaptive FIR filters are always stable. On the contrary, the stability of the filter depends critically on the algorithm for adjusting its coefficients.

Of the various FIR filter structures that we may use, the direct form and the lattice form are the ones often used in adaptive filtering applications. The direct-form FIR filter structure with adjustable coefficients $h(0), h(1), ..., h(N-1)$ is illustrated in Figure 10-1. On the other hand, the adjustable parameters in an FIR lattice structure are the reflection coefficients $\{K_n\}$.

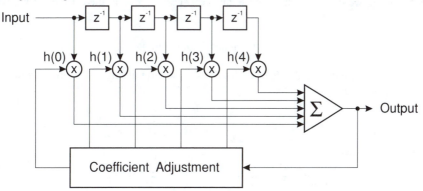

Figure 10-1: Direct-Form Adaptive FIR Filter

An important consideration in the use of an adaptive filter is the criterion for optimizing the adjustable filter parameters. The criterion must not only provide a meaningful measure of filter performance, but it must also result in a practically realizable algorithm.

One criterion that provides a good measure of performance in adaptive filtering applications is the least-squares criterion, and its counterpart in a statistical formulation of the problem, namely, the mean-square-error (MSE) criterion. The least squares (and MSE) criterion results in a quadratic performance index as a function of the filter coefficients and hence it possesses a single minimum. The resulting algorithms for adjusting the coefficients of the filter are relatively easy to implement.

In this chapter, we describe a basic algorithm, called the *least-mean- square (LMS) algorithm* to adaptively adjust the coefficients of an FIR filter. The adaptive filter structure that will be implemented is the direct-form FIR filter structure with adjustable coefficients $h(0), h(1), ..., h(N-1)$, as illustrated in Figure 10-1. After we describe the LMS algorithm, we apply it to several practical systems in which adaptive filters are employed.

10.2 LMS ALGORITHM FOR COEFFICIENT ADJUSTMENT

Suppose we have an FIR filter with adjustable coefficients $\{h(k), 0 \leq k \leq N-1\}$. Let $\{x(n)\}$ denote the input sequence to the filter and let the corresponding output be $\{y(n)\}$, where

$$y(n) = \sum_{k=0}^{N-1} h(k)x(n-k) \qquad (10-1)$$

Suppose that we also have a desired sequence $\{d(n)\}$ with which we can compare the FIR filter output. Then, we can form the error sequence $\{e(n)\}$ by taking the difference between $d(n)$ and $y(n)$. That is

$$e(n) = d(n) - y(n) \qquad (10-2)$$

The coefficients of the FIR filter will be selected to minimize the sum of squared errors. Thus, we have

$$\mathcal{E} = \sum_{n=0}^{M} e^2(n) = \sum_{n=0}^{M} \left[d(n) - \sum_{k=0}^{N-1} h(k)x(n-k) \right]^2$$

$$= \sum_{n=0}^{M} d^2(n) - 2\sum_{k=0}^{N-1} h(k)r_{dx}(k) + \sum_{k=0}^{N-1}\sum_{l=0}^{N-1} h(k)h(l)r_{xx}(k-l) \qquad (10-3)$$

where, by definition,

$$r_{dx}(k) = \sum_{n=0}^{M} d(n)x(n-k) \quad , \quad 0 \leq k \leq N-1 \qquad (10-4)$$

$$r_{xx}(k) = \sum_{n=0}^{M} x(n)x(n+k) \quad , \quad 0 \leq k \leq N-1 \qquad (10-5)$$

We call $\{r_{dx}(k)\}$ the cross-correlation between the desired output sequence $\{d(n)\}$ and the input sequence $\{x(n)\}$ and $\{r_{xx}(k)\}$ is the auto-correlation sequence of $\{x(n)\}$.

The sum of squared errors $\mathcal{E}$ is a quadratic function of the FIR filter coefficients. Consequently, the minimization of $\mathcal{E}$ with respect to the filter coefficients $\{h(k)\}$ results in a set of linear equations. By differentiating $\mathcal{E}$ with respect to each of the filter coefficients we obtain

$$\frac{\partial \mathcal{E}}{\partial h(m)} = 0 \quad , \quad m = 0, 1, ..., N-1 \qquad (10-6)$$

and, hence,

$$\sum_{k=0}^{N-1} h(k)r_{xx}(k-m) = r_{dx}(m) \quad , \quad m = 0, 1, ..., N-1 \qquad (10-7)$$

This is the set of linear equations which yield the optimum filter coefficients.

In order to solve the set of linear equations directly, we must first compute the auto-correlation sequence $\{r_{xx}(k)\}$ of the input signal and the cross-correlation sequence $\{r_{dx}(k)\}$ between the desired sequence $\{d(n)\}$ and the input sequence $\{x(n)\}$.

The LMS algorithm provides an alternative computational method for determining the optimum filter coefficients $\{h(k)\}$ without explicitly computing the correlation sequences $\{r_{xx}(k)\}$ and $\{r_{dx}(k)\}$. The algorithm is basically a recursive gradient (steepest-descent) method that finds the minimum of $\mathcal{E}$ and, thus, yields the set of optimum filter coefficients.

We begin with any arbitrary choice for the initial values of $\{h(k)\}$, say $\{h_0(k)\}$. For example, we may begin with $h_0(k) = 0, 0 \le k \le N$. Then, after each new input sample $x(n)$ enters the adaptive FIR filter, we compute the corresponding output, say $y(n)$, form the error signal $e(n) = d(n) - y(n)$ and update the filter coefficients according to the equation

$$h_n(k) \quad = \quad h_{n-1}(k) + \Delta e(n)x(n-k) \quad , \quad k = 0, 1, \ldots, N-1 \quad , \quad n = 1, 2, \ldots \qquad (10-8)$$

where Δ is called the step size parameter and $x(n-k)$ is the sample of the input signal located at the kth tap of the filter at time n. This is the LMS recursive algorithm for adjusting the filter coefficients adaptively so as to minimize the sum of squared errors $\mathcal{E}$.

The step size parameter Δ controls the rate of convergence of the algorithm to the optimum solution. A large value of Δ leads to large step size adjustments and, thus, to rapid convergence, while a small value of Δ results in slower convergence. However, if Δ is made too large the algorithm becomes unstable. To ensure stability Δ must be chosen to be in range

$$0 < \Delta < \frac{1}{10NP_x} \qquad (10-9)$$

where N is the length of the adaptive FIR filter and P_x is the power in the input signal, which can be approximated by

$$P_x \quad \approx \quad \frac{1}{M+1} \sum_{n=0}^{M} x^2(n) \quad = \quad \frac{r_{xx}(0)}{M+1} \qquad (10-10)$$

The mathematical justification of the equations (10-9) and (10-10) and the proof that the LMS algorithm leads to the solution for the optimum filter coefficients is given in more advanced treatments of adaptive filters. The interested reader may refer to the books by Haykin [24] and Proakis [21].

Below, we apply the LMS algorithm to several practical applications involving adaptive filtering.

10.3 SYSTEM IDENTIFICATION OR SYSTEM MODELING

To formulate the problem, let us refer to Figure 10-2. We have an unknown linear system that we wish to identify. The unknown system may be an all-zero (FIR) system or a pole-zero (IIR) system.   The unknown system will be approximated (modeled) by an FIR filter of length N. Both the unknown system and the FIR model are connected in parallel and are excited by the same input sequence $\{x(n)\}$. If $\{y(n)\}$ denotes the output of the model and $\{d(n)\}$ denotes the output of the unknown system, the error sequence is $\{e(n) = d(n) - y(n)\}$. If we minimize the sum of squared errors, we obtain the same set of linear equations as in (10-7). Therefore, the LMS algorithm given by (10-8) may be used to adapt the coefficients of the FIR model so that its output approximates the output of the unknown system.

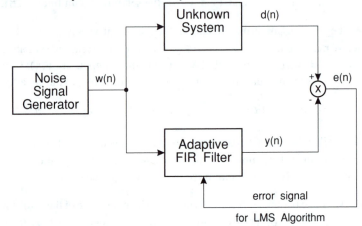

Figure 10-2: Block Diagram of System Identification or System Modeling Problem

Experiment in System Identification

There are three basic software modules that are needed to perform this experiment.

1. A noise signal generator that generates a sequence of random numbers with zero mean value. For example, we may generate a sequence of uniformly distributed random numbers over the interval $(-a, a)$. Such a sequence of uniformly distributed numbers has an average value of zero and a variance of $a^2/12$. This signal sequence, call it $\{x(n)\}$, will be used as the input to the unknown system and the adaptive FIR model. In this case, the input signal $\{x(n)\}$ has power $P_x = a^2/12$.

2. An unknown system module which may be selected as an IIR filter and implemented by its difference equation. For example, we may select an IIR filter specified by the (two-pole, two-zero) difference equation

$$d(n) \;=\; a_1 d(n-1) + a_2 d(n-2) + x(n) + b_1 x(n-1) + b_2 x(n-2) \qquad (10-11)$$

where the choice parameters (a_1, a_2) determine the positions of the poles and (b_1, b_2) determine the positions of the zeros of the filter. These parameters are input variables to the program.

3. An adaptive FIR filter module where the FIR filter has N tap coefficients that are adjusted by means of the LMS algorithm. The length N of the filter is an input variable to the program.

The three modules are configured as shown in Figure 10-2. From this experiment we can determine how closely the impulse response of the FIR model approximates the impulse response of the unknown system after the LMS algorithm has converged.

To monitor the convergence rate of the LMS algorithm, we may compute a short-term average of the squared error $e^2(n)$ and display it. That is, we may compute

$$\text{ASE}(m) \;=\; \frac{1}{K} \sum_{k=n+1}^{n+K} e^2(n) \qquad\qquad (10-12)$$

where $m = n/K = 1, 2, \ldots$. The averaging interval K may be selected in the range $10 \le K \le 25$. The effect of the choice of the step size parameter Δ on the convergence rate of the LMS algorithm may be observed by monitoring the $\text{ASE}(m)$.

A Flowchart of the system identification program is shown in Figure 10-3. Besides the main part of the program, we have also included, as an aside, the computation of the impulse response of the unknown system, which can be obtained by exciting the system with a unit sample sequence $\delta(n)$. This actual impulse response can be compared with that of the FIR model after convergence of the LMS algorithm. The two impulse responses can be displayed for the purpose of comparison.

10.4 SUPPRESSION OF NARROWBAND INTERFERENCE IN A WIDEBAND SIGNAL

Let us assume that we have a signal sequence $\{x(n)\}$ that consists of a desired wideband signal sequence, say $\{w(n)\}$ corrupted by an additive narrowband interference sequence $\{s(n)\}$. The two sequences are uncorrelated. This problem arises in digital communications and in signal detection, where the desired signal sequence $\{w(n)\}$ is a spread-spectrum signal while the narrowband interference represents a signal from another user of the frequency band or some intentional interference from a jammer who is trying to disrupt the communication or detection system.

From a filtering point of view, our objective is to design a filter that suppresses the narrowband interference. In effect, such a filter should place a notch in the frequency band occupied by the interference. In practice, however, the frequency band of the interference might be unknown. Moreover, the frequency band of the interference may vary slowly in time.

The narrowband characteristics of the interference allow us to estimate $s(n)$ from past samples of the sequence $x(n) = s(n) + w(n)$ and to subtract the estimate from $x(n)$. Since the bandwidth of $\{s(n)\}$ is narrow compared to the bandwidth of $\{w(n)\}$, the samples of $\{s(n)\}$ are highly correlated. On the other hand, the wideband sequence $\{w(n)\}$ has a relatively narrow correlation.

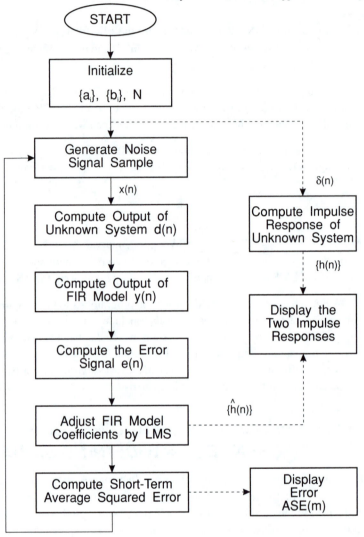

Figure 10-3: Flowchart of System Identification Program

The general configuration of the interference suppression system is shown in Figure 10-4. The signal $x(n)$ is delayed by D samples, where the delay D is chosen sufficiently large so that the wideband signal components $w(n)$ and $w(n-D)$, which are contained in $x(n)$ and $x(n-D)$, respectively, are uncorrelated. The output of the adaptive FIR filter is the estimate

$$\hat{s}(n) = \sum_{k=0}^{N-1} h(k)x(n-k-D) \qquad (10-13)$$

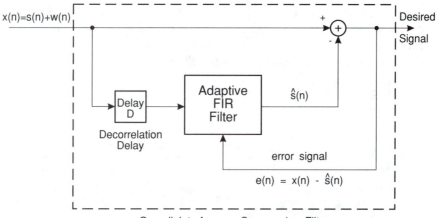

Overall Interference Suppression Filter

Figure 10-4: Adaptive Filter for Estimating and Suppressing a Narrowband Interference in a Wideband Signal

The error signal that is used in optimizing the FIR filter coefficients is $e(n) = x(n) - \hat{s}(n)$. The minimization of the sum of squared errors again leads to a set of linear equations for determining the optimum coefficients. Due to the delay D, the LMS algorithm for adjusting the coefficients recursively becomes

$$h_n(k) \;=\; h_{n-1}(k) + \Delta e(n)x(n-k-D) \quad , \quad \begin{matrix} k = 0, 1, ..., N-1 \\ n = 1, 2, ... \end{matrix} \qquad (10-14)$$

Experiment on Suppression of Sinusoidal Interference

There are three basic software modules required to perform this experiment.

1. A noise signal generator module that generates a wideband sequence $\{w(n)\}$ of random numbers with zero mean value. In particular, we may generate a sequence of uniformly distributed random numbers as previously described in the experiment on system identification. The signal power is denoted as P_w.

2. A sinusoidal signal generator module that generates a sine wave sequence $s(n) = A \sin \omega_0 n$, where $0 < \omega_0 < \pi$ and A is the signal amplitude. The power of the sinusoidal sequence is denoted as P_s.

3. An adaptive FIR filter module where the FIR filter has N tap coefficients that are adjusted by the LMS algorithm. The length N of the filter is an input variable to the program.

The three modules are configured as shown in Figure 10-5. In this experiment the delay $D = 1$ is sufficient, since the sequence $\{w(n)\}$ is a white noise (spectrally flat or uncorrelated) sequence. The objective is to adapt the FIR filter coefficients and then to investigate the characteristics of the adaptive filter.

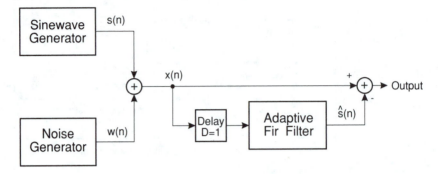

Figure 10-5: Configuration of Modules for Experiment on Interference Suppression

It is interesting to select the interference signal to be much stronger than the desired signal $w(n)$, for example, $P_s = 10P_w$. Note that the power P_x required in selecting the stepsize parameter in the LMS algorithm is $P_x = P_s + P_w$. The frequency response characteristic $H(\omega)$ of the adaptive FIR filter with coefficients $\{h(k)\}$ should exhibit a resonant peak at the frequency of the interference. The frequency response of the interference suppression filter is $H_s(\omega) = 1 - H(\omega)$, which should then exhibit a notch at the frequency of the interference.

A flowchart for this experiment is shown in Figure 10-6. It is interesting to display the sequences $\{w(n)\}$, $\{s(n)\}$, and $\{x(n)\}$. It is also interesting to display the frequency responses $H(\omega)$ and $H_s(\omega)$ after the LMS algorithm has converged. The short-time average squared error ASE(m), defined by (10-12) may be used to monitor the convergence characteristics of the LMS algorithm. The effect of the length of the adaptive filter on the quality of the estimate should be investigated.

The experiment may be generalized by adding a second sinusoid of a different frequency. Then, $H(\omega)$ should exhibit two resonant peaks, provided the frequencies are sufficiently separated. Investigate the effect of the filter length N on the resolution of two closely spaced sinusoids.

10.5 ADAPTIVE LINE ENHANCEMENT

In the preceding section we described a method for suppressing a strong narrowband interference from a wideband signal. An adaptive line enhancer (ALE) has the same configuration as the interference suppression filter in Figure 10-4, except that the objective is different.

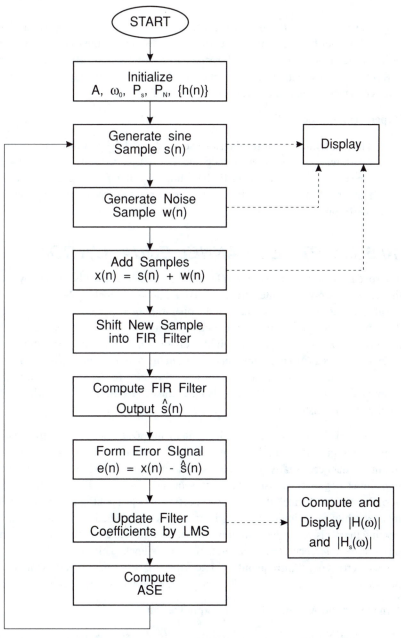

Figure 10-6: Flowchart for Experiment on Suppression of Narrowband Interference

In the adaptive line enhancer, $\{s(n)\}$ is the desired signal and $\{w(n)\}$ represents a wideband noise component that masks $\{s(n)\}$. The desired signal $\{s(n)\}$ may be a spectral line (a pure sinusoid) or a relatively narrowband signal. Usually, the power in the wideband signal is greater than that in the narrowband signal, i.e., $P_w > P_s$. It is apparent that the ALE is a self-tuning filter that has a peak in its frequency response at the frequency of the input sinusoid or in the frequency band occupied by the narrowband signal. By having a narrow bandwidth FIR filter, the noise outside the frequency band of the signal is suppressed and, thus, the spectral line is enhanced in amplitude relative to the noise power in $\{w(n)\}$.

Experiment on the ALE

This experiment requires the same software modules as those used in the experiment on interference suppression. Hence, the description given in Section 10.4 applies directly. One change is that in the ALE, the condition is that $P_w > P_s$. Secondly, the output signal from the ALE is $\{s(n)\}$. Repeat the experiment described in the previous section under these conditions.

10.6 ADAPTIVE CHANNEL EQUALIZATION

The speed of data transmission over telephone channels is usually limited by channel distortion that causes intersymbol interference (ISI). At data rates below 2400 bits the ISI is relatively small and is usually not a problem in the operation of a modem. However, at data rates above 2400 bits, an adaptive equalizer is employed in the modem to compensate for the channel distortion and, thus, to allow for highly reliable high speed data transmission. In telephone channels, filters are used throughout the system to separate signals in different frequency bands. These filters cause amplitude and phase distortion. The adaptive equalizer is basically an adaptive FIR filter with coefficients that are adjusted by means of the LMS algorithm to correct for the channel distortion.

A block diagram showing the basic elements of a modem transmitting data over a channel is given in Figure 10-7. Initially, the equalizer coefficients are adjusted by transmitting a short training sequence, usually less than 1 second in duration. After the short training period, the transmitter begins to transmit the data sequence $\{a(n)\}$. To track the possible slow time variations in the channel, the equalizer coefficients must continue to be adjusted in an adaptive manner while receiving data. This is usually accomplished, as illustrated in Figure 10-7, by treating the decisions at the output of the decision device as correct, and using the decisions in place of the reference $d(n)$ to generate the error signal. This approach works quite well when decision errors occur infrequently, e.g. less than one error in 100 data symbols. The occasional decision errors cause only a small misadjustment in the equalizer coefficients.

Experiment on Adaptive Channel Equalization

The objective of this experiment is to investigate the performance of an adaptive equalizer for data transmission over a channel that causes intersymbol interference. The basic configuration of the system to be simulated is shown in Figure 10-8. As we observe, five basic software modules are required. Note that we have avoided carrier modulation and demodulation, which is required in a telephone channel modem. This is done in order to simplify the simulation program. However, all processing involves complex arithmetic operations.

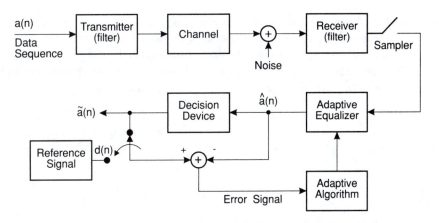

Figure 10-7: Application of Adaptive Filtering to Adaptive Channel Equalization

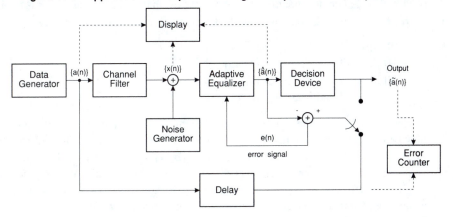

Figure 10-8: Experiment for Investigating the performance of an Adaptive Equalizer in the presence of Channel Distortion

The five software modules are as follows:

1. The data generator module is used to generate a sequence of complex-valued information symbols $a(n)$. In particular, let us employ four equally probable symbols $s + js, s - js, -s + js$ and $-s - js$, where s is a scale factor that may be set to $s = 1$, or it can be a parameter selected by the programmer.

2. The channel filter module is an FIR filter with coefficients $\{c(n), 0 \leq n \leq K - 1\}$ which simulate the channel distortion. For distortionless transmission, we set $c(0) = 1$ and $c(n) = 0$ for $1 \leq n \leq K - 1$. The length K of the filter is a parameter that is selected by the programmer.

3. The noise generator module is used to generate additive noise that is usually present in any digital communication system. If we are modeling noise that is generated by electronic devices, the noise distribution should be Gaussian with zero mean. Such noise can be generated by adding six to twelve uniformly distributed random numbers and scaling the sum to obtain the desired noise power.

4. The adaptive equalizer module is an FIR filter with tap coefficients $\{h(k), 0 \leq k \leq N-1\}$ which are adjusted by the LMS algorithm. However, due to the use of complex arithmetic, the recursive equation in the LMS algorithm is slightly modified to

$$h_n(k) \;=\; h_{n-1}(k) \;+\; \Delta e(n) x^*(n-k) \qquad\qquad (10-15)$$

where the asterisk denotes the complex conjugate.

5. The decision device module which takes the estimate $\hat{a}(n)$ and quantizes it to one of the four possible signal points on the basis of the following decision rule:

$\Re[\hat{a}(n)] > 1$ and $\Im[\hat{a}(n)] > 1 \;\rightarrow\; 1+j$
$\Re[\hat{a}(n)] > 1$ and $\Im[\hat{a}(n)] < 1 \;\rightarrow\; 1-j$
$\Re[\hat{a}(n)] < 1$ and $\Im[\hat{a}(n)] > 1 \;\rightarrow\; -1+j$
$\Re[\hat{a}(n)] < 1$ and $\Im[\hat{a}(n)] < 1 \;\rightarrow\; -1-j$

The effectiveness of the equalizer in suppressing the ISI introduced by the channel filter may be seen by displaying the following relevant sequences in a two-dimensional (real-imaginary) display. The data generator output $\{a(n)\}$ should consist of four points with values $\pm 1 \pm j$. The effect of channel distortion and additive noise may be viewed by displaying the sequence $\{x(n)\}$ at the input to the equalizer. The effectiveness of the adaptive equalizer may be assessed by displaying its output $\{\hat{a}(n)\}$ after convergence of its coefficients. The short-time average squared error $ASE(n)$ may also be used to monitor the convergence characteristics of the LMS algorithm. Note that a delay must be introduced into the output of the data generator to compensate for the delays that the signal encounters due to the Channel Filter and the Adaptive Equalizer. For example, this delay may be set to the largest integer closest to $(N+K)/2$. Finally, an error counter may be used to count the number of symbol errors in the received data sequence and the ratio for the number of errors to the total number of symbols (error rate) may be displayed. The error rate may be varied by changing the level of the ISI and the level of the additive noise.

It is suggested that simulations be performed for the following three channel conditions:

(a)	No ISI:	$c(0)=1$,			$c(n)=0$,	$1 \leq n \leq K\text{-}1$
(b)	Mild ISI:	$c(0)=1$,	$c(1)=0.2$,	$c(2)=-0.2$,	$c(n)=0$,	$3 \leq n \leq K\text{-}1$
(c)	Strong ISI:	$c(0)=1$.	$c(1)=0.5$,	$c(2)=0.5$,	$c(n)=0$,	$3 \leq n \leq K\text{-}1$

The measured error rate may be plotted as a function of the signal-to-noise ratio (SNR) at the input to the equalizer, where SNR is defined as P_s/P_n, where P_s is the signal power, given as $P_s = s^2$ and P_n is the noise power of the sequence at the output of the noise generator.

10.7 ADAPTIVE ECHO CANCELLATION

It is well known that modems are used in the transmission of data over telephone channels. Shown in Figure 10-9 is a block diagram of a communication system in which two terminals, labeled A and B, transmit data by using modems A and B to interface to a telephone channel. As shown, a digital sequence $a(n)$ is transmitted from terminal A to terminal B while a digital sequence $b(n)$ is transmitted from terminal B to A. This simultaneous transmission in both directions is called *full-duplex transmission*.

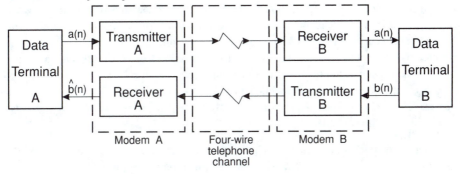

Figure 10-9: Full-Duplex Data Transmission over Telephone Channels

When a subscriber leases a private line from a telephone company for the purpose of transmitting data between terminals A and B, the telephone line provided is a four-wire line, which is equivalent to having two dedicated telephone (two-wire) channels, one (pair of wires) for transmitting data in one direction and one (pair of wires) for receiving data from the other direction. In such a case the two transmission paths are isolated, and consequently, there is no "crosstalk" or mutual interference between the two signal paths. Channel distortion is compensated by use of an adaptive equalizer, as described above, at the receiver of each modem.

The major problem with the system shown in Figure 10-9 is the cost of leasing a four-wire telephone channel. If the volume of traffic is high and the telephone channel is used either continuously or frequently, as in banking transactions systems or airline reservation systems, the system pictured in Figure 10-9 is cost-effective. Otherwise, it is not.

An alternative solution for low-volume, infrequent transmission of data is to use the dial-up switched telephone network. In this case, the local communication link between the subscriber and the local central telephone office is a two-wire line, called the *local loop*. At the central office, the subscriber two-wire line is connected to the main four-wire telephone channels that interconnect different central offices, called *trunk lines*, by a device called a *hybrid*. By using transformer coupling, the hybrid is tuned to provide isolation between the transmit and receiver channels in full-duplex operation. However, due to impedance mismatch between the hybrid and the telephone channel, the level of isolation is often insufficient, and consequently, some of the signal on the transmit side leaks back and corrupts the signal on the receiver side, causing an "echo" that is often heard in voice communications over telephone channels.

To mitigate the echoes in voice transmission, the telephone companies employ a device called an *echo suppressor*. In data transmission, the solution is to use an *echo canceller* within each modem. The echo cancellers are implemented as adaptive FIR filters with automatically adjustable coefficients.

With the use of hybrids to couple a two-wire channel to a four-wire channel and echo cancellers at each modem to estimate and subtract out the echoes, the data communication system for the dial-up switched network takes the form shown in Figure 10-10. A hybrid is needed at each modem to isolate the transmitter from the receiver and to couple to the two-wire local loop. Hybrid A is physically located at the central office of subscriber A, while Hybrid B is located at the central office to which subscriber B is connected. The two central offices are connected by a four-wire line, one pair used for transmission from A to B and the other pair is used for transmission in the reverse direction, from B to A. An echo at terminal A due to the hybrid A is called a *near-end-echo*, while an echo at terminal A due to the hybrid B is termed a *far-end-echo*. Both types of echos are usually present in data transmission and must be removed by the echo canceller.

Suppose that we neglect the channel distortion for purposes of this discussion, and let us deal with echoes only. The signal received at modem A may be expressed as

$$S_{RA}(t) = A_1 S_B(t) + A_2 S_A(t - d_1) + A_3 S_A(t - d_2) \tag{10-16}$$

where $S_B(t)$ is the desired signal to be demodulated at modem A, $S_A(t - d_1)$ is the near-end echo due to hybrid A, $S_A(t - d_2)$ is the far-end echo due to hybrid B, and $A_i, i = 1, 2, 3$ are the corresponding amplitudes of the three signal components and (d_1, d_2). are the delays associated with the echo components. A further disturbance that corrupts the received signal is additive noise, so that the received signal at modem A is

$$r_A(t) = S_{RA}(t) + w(t) \tag{10-17}$$

where $w(t)$ represents the additive noise process.

The adaptive echo canceller attempts to estimate adaptively the two-echo components. If its coefficients are $h(n), n = 0, 1, \ldots, M - 1$, its output is

$$\hat{s}_A(n) = \sum_{k=0}^{M-1} h(k) a(n - k) \tag{10-18}$$

which is an estimate of the echo signal components. This estimate is subtracted from the sampled received signal and the resulting error signal can be minimized in the least-squares sense to adjust optimally the coefficients of the echo canceller.

Experiment on Echo Cancellation

The objective of this experiment is to investigate the effectiveness of echo cancellation in a data communication system. The basic configuration of the system to be simulated is shown in Figure 10-11. The following basic software modules are used in the experiment.

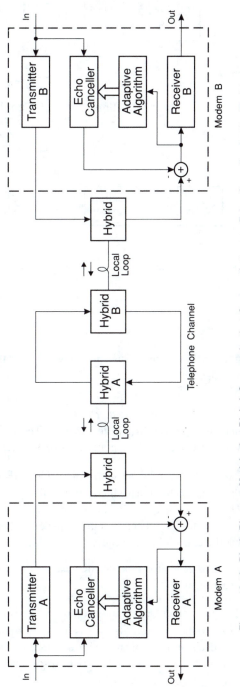

Figure 10-10: Block Diagram Model of a Digital Communication System that uses Echo Canceller in the Modems.

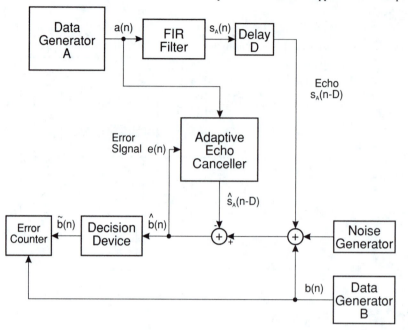

Figure 10-11: Block Diagram of System for Investigating the Performance of an Adaptive Echo Canceller

1. Two separate data generator modules are used to generate two binary (±1's) information sequences $\{a(n)\}$ and $\{b(n)\}$. The sequence $\{b(n)\}$ is the desired sequence which is detected. The sequence $\{a(n)\}$ serves as the interfering sequences.

2. Two FIR filter modules are needed. One FIR filter of length M is used to generate the (single) echo signal which serves as the interference in the detection of the sequence $\{b(n)\}$. The second FIR filter is the adaptive echo canceller which has length $N > M$, say $N = M + D$, where D is the echo delay. The LMS algorithm is used to adapt the coefficients of the echo canceller.

3. A noise generator module identical to the one used in the adaptive equalization experiment.

4. The decision-device module compares the estimate $\hat{b}(n)$ with the threshold zero. If $\hat{b}(n) > 0$ the decision $\tilde{b}(n) = 1$ is made. If $\hat{b}(n) < 0$, the decision $\tilde{b}(n) = -1$ is made. The decision is compared with the transmitted bit $b(n)$ and an error is counted if $\tilde{b}(n) \neq b(n)$. The error rate may be displayed for different signal-to-noise ratios.

In practice the echo signal is usually of higher power than the desired signal $b(n)$. For this reason, the step size parameter Δ in the LMS algorithm may have to be made much smaller than the value given by (10-9) in order to obtain sufficient suppression of the echo. One strategy is to begin with the value given by (10-9) and, then, to reduce this value by factors of two every one to two hundred iterations. The convergence rate of the LMS algorithm may be monitored by computing and displaying the ASE(n).

After the coefficients of the echo canceller have converged, it is interesting to compare its output $\{\hat{s}_A(n-D)\}$ with the exact value of the echo signal $\{s_A(n-D)\}$. These two signal sequences may also be displayed for purposes of comparison.

10.8 SUMMARY

In this chapter we introduced the reader to the theory and implementation of adaptive FIR filters with applications to system identification, interference suppression, narrowband frequency enhancement, adaptive equalization, and echo cancellation. Experiments were formulated involving these applications of adaptive filtering which may be implemented on the ADSP-2100 family of microcomputers.

REFERENCES

1. Analog Devices, Inc., *ADSP-2101/2102 User's Manual, First Edition*, Analog Devices, Inc., Norwood, 1990.

2. Analog Devices, Inc., *ADSP-2101 Cross-Software Manual, First Edition*, Analog Devices, Inc., Norwood, 1989.

3. Analog Devices, Inc., *ADSP-2101 EZ-ICE™ Manual, First Edition*, Analog Devices, Inc., Norwood, 1990.

4. Analog Devices, Inc., *ADSP-2101 EZ-Lab™ Manual, First Edition*, Analog Devices, Inc., Norwood, MA 1990.

5. R. J. Higgins, *Digital Signal Processing in VLSI*, Prentice Hall, Englewood Cliffs, NJ, 1990.

6. Analog Devices, Inc., *Digital Signal Processing Applications using the ADSP-2100 Family*, Prentice Hall, Englewood Cliffs, NJ, 1990.

7. D. E. Knuth, *The Art of Computer Programming: Volume 2 / Seminumerical Algorithms, Second Edition*, Addison-Wesley Publishing Company, Reading, MA, 1969.

8. J. G. Proakis and D. G. Manolakis, *Introduction to Digital Signal Processing*, Macmillan, NY, 1988.

9. T. W. Parks and J. H. McClellan, "A Program for the Design of Linear Phase Finite Impulse Response Digital Filters," *IEEE Trans. Audio and Electroacoustics,* Vol. AU-20, pp. 195-199, August 1972.

10. A. B. Carlson, *Communication Systems,* McGraw Hill, Inc., New York, NY, 1975.

11. L. R. Rabiner and B. Gold, *Theory and Applications in Digital Signal Processing,* Prentice Hall, Englewood Cliffs, NJ, 1975.

12. J. Durbin, "Efficient Estimation of Parameters in Moving-Average Models," *Biometrika,* vol. 46, parts 1 and 2, pp. 306-316, 1959.

13. J. L. Flanagan, et al., "Speech Coding," *IEEE Trans. Commun.* vol. COM-27, pp. 710-736, April, 1979.

14. D. A. George, R. R. Bowen, and J.R. Storey, "An Adaptive Decision-Feedback Equalizer," *IEEE Trans. Commun. Tech.,* vol. COM-19, pp. 281-293, June, 1971.

15. J. A. Greefles, "A Digitally Companded Delta Modulation Modem for Speech Transmission," *Proce. IEEE Int. Conf. on Communications*, pp. 7.33 - 7.48, June, 1970.

16. F. M. Hsu and A. A. Giordano "Digital Whitening Techniques for Improving Spread Spectrum Communications Performance in the Presence of Narrowband Jamming and Interference," *IEEE Trans. Commun.*, vol. COM-26, pp. 209-216, February, 1978.

17. N.S. Jayant, "Adaptive Delta Modulation with a One-Bit Memory," *Bell Syst. Tech. J.*, pp. 321-342, March, 1970.

18. N. S. Jayant, "Digital Coding of Speech Waveforms: PCM, DPCM, and DM Quantizers," *Proc. IEEE*, vol. 62, pp. 611-632, May, 1974.

19. J. W. Ketchum and J. G. Proakis, "Adaptive Algorithms for Estimation and Suppression of Narrowband Interference in PN Spread-Spectrum Systems," *IEEE Trans. Communications*, vol. COM-30, pp. 913-923, May, 1982.

20. N. Levinson, "The Wiener RMS (Root Mean Square) Error Criterion in Filter Design and Prediction," *J. Math. Phys.*, vol. 25, pp. 261-278, 1947.

21. J. G. Proakis, *Digital Communications*, McGraw-Hill, New York, NY, 1989.

22. B. Widrow, P. Manley and L. J. Griffiths, "Adaptive Antenna Systems," *Proc. IEEE*, vol. 55, pp. 2143-2159, December, 1967.

23. Widrow, et al. "Adaptive Noise Cancelling: Principles and Applications," *Proc. IEEE*, vol. 63, pp. 1692-1716, December, 1975.

24. Haykin, S., *Adaptive Filter Theory*, Prentice Hall, Englewood Cliffs, NJ, 1986.

INDEX